THE
SADTLER
GUIDE TO THE
NMR SPECTRA
OF POLYMERS

By
W. W. SIMONS
NMR Spectroscopist
Sadtler Research Laboratories, Inc.

and

M. ZANGER
Professor of Chemistry
Philadelphia College of Pharmacy and Science

1973

Sadtler Research Laboratories, Inc.
Subsidiary of Block Engineering, Inc.
3316 Spring Garden Street
Philadelphia, Pa. 19104

Published by

Sadtler Research Laboratories, Inc.

Subsidiary of Block Engineering, Inc.

3316 Spring Garden Street

Philadelphia, Pennsylvania 19104

ISBN 0-8456-0002-8

Library of Congress Catalog Card Number: 73-90432

Co-published for exclusive distribution in Europe by

Heyden & Son Ltd.

Spectrum House, Alderton Crescent

London N.W. 4, England

ISBN 0-85501-098-3

Printed in the United States of America

INTRODUCTION

NUCLEAR MAGNETIC RESONANCE SPECTROSCOPY has firmly established itself as one of the primary analytical tools available to chemists today. Its analytical potential has not, as yet, been fully realized and new applications of the method are still being uncovered. NMR is primarily a structure-determining tool and nowhere has it found greater utility than in the elucidation of complex organic molecules. One class of compounds where its impact has been impressive is in the study of polymers.

NMR can be used to study polymers in three basic ways:

(1) It can be used for qualitatively identifying polymeric types.

(2) It can be used to identify polymer mixtures and give a reasonable estimate of the per cent composition of the individual components.

(3) It can be used to determine the microtacticity of portions of the polymer chain; that is, it can be used to identify the configurational sequence of short segments of the polymer backbone.

The main purpose of this book is to enable the user to readily characterize specific polymers. At first, this might appear to be an easy task, but due to the wide variety of polymers extant and the large number of variations of the basic polymer types, the problem is found to be complex. Other analytical methods (e.g. Infrared Spec., Thermal Gravimetric Analysis, etc.) also provide useful information about polymers but of a type unlike that provided by NMR. Used in conjunction with NMR, these other methods provide a broad analytical capability for the elucidation of the molecular structure of polymers, copolymers and resins.

A second objective of this volume is to illustrate the analysis of copolymers and polymer mixtures with an approach to quantitation. In many cases, it will be seen, a fairly accurate quantitative analysis may be carried out in a matter of minutes. The last application of the NMR method, the determination of tacticity, will be discussed only briefly since the fine structure of polymers is best studied with instruments of higher resolution (e.g. 220 MHz). Tacticity will, however, be mentioned in the text wherever its effects can be observed in a specific spectrum.

In this volume we seek mainly to illustrate the spectra of a variety of common polymers and copolymers and to indicate the features which make them unique. We will attempt to do this in a systematic fashion, using a minimum of organic and polymer chemistry, so that those whose background is mainly analytical, can utilize this book in its entirety.

TABLE OF CONTENTS

NMR THEORY AND DEFINITIONS

It is not the intent of this volume to give a complete description of the NMR method, its theory and applications. This has already been done in a variety of other texts. Rather, the usefulness of this text is in its specialized approach to the description and analysis of the NMR spectra of Polymers.

These spectra may be utilized for identification purposes by individuals with a minimum of training just as infrared spectra are used by the majority of chemists.

It would be useful at this point to describe briefly some of the terms and concepts which are important to the application of this method to the analysis of polymers.

I. <u>NMR THEORY</u>: Certain atomic nuclei behave as if they are small magnets. When these nuclei are placed between the poles of a magnet, they will align themselves either with or against the applied magnetic field (H_O). The nuclei now exist in two (or more) orientations and it is possible to observe energy transitions between these high and low energy states. This is done by irradiating the sample with energy in the radio-frequency range. In order for the low energy nuclei to absorb energy, the fundamental NMR equation (Eq. 1) must be obeyed:

$$\nu \, / \, H_O \; = \; \gamma \, / \, 2\pi$$

where: ν = irradiating frequency

H_O = strength of applied field

Equation 1

γ = magnetogyric constant

Briefly, this equation states that for every value of the applied field, there is only one frequency at which the nuclei can absorb energy. When this absorption of energy occurs, the nuclei are said to have been brought into resonance, hence the origin of the term <u>NUCLEAR MAGNETIC RESONANCE.</u> This absorption of energy can be detected and displayed on an oscilloscope or chart recorder as a "peak".

Although many nuclei possess magnetic properties, for a variety of reasons only certain of them are susceptible to the NMR method. Among the nuclei which are best studied by this technique are ^{1}H, ^{19}F, ^{13}C and ^{31}P. Because the magnetogyric constant (γ) for each nucleus is unique, each will resonate at a different frequency (ν) at a constant applied field (H_O). Conversely, if the frequency is kept constant, each nucleus will require a different applied field in order to satisfy the equation. Ordinarily, only one type of nucleus can be studied at one time. Since protons are by far the most commonly observed nuclei, especially in polymer NMR, no further mention of heteronuclear magnetic resonance will be made.

II. THE CHEMICAL SHIFT: Originally it was believed that all protons regardless of their chemical environment would resonate at the same frequency. In actuality it was observed that protons in different structural positions in a molecule came into resonance at different frequencies. The difference between the frequencies (or fields) at which two different nuclei come into resonance is termed the CHEMICAL SHIFT.

The chemical shift arises because protons in different structural environments are shielded to differing extents by the extranuclear electrons. Thus individual protons may experience local magnetic fields which are somewhat less than the total applied magnetic field. This SHIELDING effect also varies according to certain time dependent effects such as the proximity of other molecules in the sample solution or their orientation in the magnetic field. These effects can often be averaged out by using non-viscous solutions and by spinning the sample. What cannot be averaged out is the relative position in a molecule in which a proton is situated. Thus the chemical shift may be regarded as a sensitive measure of the electron density around each type of proton.

An NMR spectrometer is capable of "seeing" different proton groups based mainly on differences in their chemical shifts. Because the protons in tetramethylsilane (TMS) are among the most shielded protons found in organic molecules, a small amount of TMS is incorporated in all sample solutions as an internal reference and chemical shifts are measured in relation to the resonance of this material. Such measurements are usually expressed in the (δ) delta scale in units of parts per million (ppm). TMS is arbitrarily designated as 0.0 ppm. and most other proton groups occur DOWNFIELD from TMS. On a 60 MHz instrument, each 60 Hz of separation from the TMS band is designated as 1.0 ppm. The typical maximum range for protons is about 12 - 15 ppm. The NMR spectroscopist identifies proton groups, with designations such as methyl (CH_3-), methylene ($-CH_2-$), methine ($-CH-$), olefinic ($C=C-H$), aromatic ($\emptyset-H$), aldehydic ($-C(=O)-H$), etc. With a little practice, the chemist will soon learn the range of chemical shifts for each proton type but for more accurate assignments and for predicting the shifts of the protons in a specific molecule, many good CHEMICAL SHIFT TABLES are available (see page 7). In a spectrum display, TMS (0.0 ppm.) is arbitrarily positioned at the far right of the chart and all common peaks occur at varying distances to the left. NMR terminology defines a peak which is at the far left or to the left of another peak as DESHIELDED, DOWNFIELD or at LOW FIELD. Peaks at or toward the right (near TMS) are termed SHIELDED, UPFIELD, or at HIGH FIELD.

The same substituent may have a different effect on aliphatic protons than it has on aromatic protons. A methoxy group, for instance, is a strongly deshielding group when attached to an aliphatic molecule (inductive effect) but is a strongly shielding group when attached to an aromatic ring (resonance effect).

When two or more protons have identical chemical shifts they are said to be EQUIVALENT. This equivalence may be of three types, CHEMICAL, MAGNETIC or ACCIDENTAL EQUIVALENCE. For this discussion we need not distinguish between these types but an understanding of the differences between these forms is vital to most detailed structure determinations.

2

Commercial NMR spectrometers are capable of scanning a spectrum from 0.0 ppm to either 8.33 or 10.0 ppm. Since some protons may appear outside this range, most instruments are capable of "offsetting" the whole spectrum until the missing peaks are displayed. An OFFSET peak in a spectrum is usually recorded above the normal spectrum baseline and the extent of the offset (e.g. 300 Hz, 5 ppm., etc.) is indicated.

All protons which are attached to carbon atoms possess chemical shifts within fairly narrow ranges. However, those protons which are attached to the heteroatoms, nitrogen, oxygen and sulfur, do not have fixed chemical shift ranges and are called EXCHANGEABLE PROTONS. The chemical shifts of exchangeable protons are very susceptible to changes in solvent, temperature, concentration, and pH. In addition, the appearance of these peaks can vary from a sharp spike to such a broad diffuse band that it can be lost in the baseline. The main reason for the unpredictability of these peaks is that they are very labile and can exchange quite easily with other labile protons (e.g. those in H_2O). This type of proton exchange can readily be demonstrated by adding a drop of deuterium oxide (D_2O) to a water immisible solution of a compound possessing exchangeable protons. Exchange at the liquid-liquid interface is usually rapid enough to replace all of the exchangeable protons by deuterium. Since the resonance of deuterium does not appear in a proton spectrum, the result will be the loss of the signal originally produced by the exchangeable protons.

III. INTEGRATION: An important feature of NMR spectra is that the area of each band is proportional to the relative number of hydrogens contained in it. When the instrument is operated in the integral mode, the spectrum appears as a series of steps rather than peaks, and the amplitude of each step is proportional to the peak area.

The integral can thus be used to accurately determine the relative number of the various types of protons within a molecule (e.g. CH_3-CH_2-OH, 3:2:1 ratio). It can also be used to quantitatively analyze mixtures of compounds. If a separate peak in the NMR spectrum can be assigned solely to each of the components, then by measuring the integrals of these peaks, it is possible to calculate the relative number of moles of each component. Both of these applications of the integration find enormous utility in the analysis of polymers.

IV. SPIN-SPIN COUPLING: Those nuclei which can be studied by NMR can themselves act like small magnets and therefore can either add or subtract the small magnetic field which they generate, to the overall field which is experienced by a neighboring nucleus (proton). The effect of these contributions is that the signal generated by an individual proton group is split into a MULTIPLET. This phenomenon is termed spin-spin coupling or "splitting". In the case of aliphatic groups, the resulting multiplicity can often be predicted by application of the (N + 1) RULE. Simply stated, this rule predicts that the band produced by a given proton group (methyl, methylene, methine, etc.) will be split into (N + 1) peaks, where (N) represents the number of protons on adjacent carbon atoms. Applying this rule allows the chemist to rapidly and unambiguously identify various structural fragments such as ethyl, n-propyl, isopropyl, n-butyl, sec-butyl, isobutyl and tertiary butyl groups, all of which appear as unique splitting patterns.

The separation in Hertz between the peaks of a spin-spin coupling multiplet is termed the <u>COUPLING CONSTANT</u> and is represented by the letter (J). The magnitude of J is often indicative of certain structural features in the molecule (e.g. conformation, cis/trans isomerism, etc.) Tables of coupling constants are available which can provide much useful information for the spectroscopist.

The following list contains some of the multiplets which are commonly encountered in polymer NMR spectra.

<u>Singlet:</u> A single resonance peak arises from an isolated, uncoupled proton group. Isolated methyl groups, disubstituted methanes and t-butyl groups appear as sharp singlets.

<u>Doublet:</u> A doublet consists of a pair of peaks of approximately the same height. The presence of a doublet indicates that the band arises from a proton group which is coupled to one other proton (or other nucleus with a spin of I). The separation between the peaks of the doublet (J value) is not significantly affected by solvent changes.

<u>Triplet:</u> Triplet multiplicity is observed as three equally spaced peaks with intensities in the ratio of 1:2:1. A triplet originates from the coupling of a proton group with two other nuclei. Each proton produces a doublet and the doublets overlap to produce the center peak of the triplet.

<u>Quartet</u>: A quartet arises from coupling to three other nuclei by the same coupling constant. The three doublets overlap to form four evenly spaced peaks with relative intensities of 1:3:3:1.

<u>Pentent, Sextet, Heptet:</u> These multiplets arise from coupling with four, five and six nuclei respectively. The outer peaks become progressively smaller in comparison to those in the center of the pattern and are often lost in the baseline noise. The relative intensities of a pentet are 1:4:6:4:1, a sextet are 1:5:10:10:5:1 and those of a heptet are 1:6:15:20:15:6:1.

Multiplets which clearly exhibit predictable splitting behavior such as this, are termed <u>FIRST ORDER.</u> In aliphatic compounds, if the chemical shift differences between coupled proton groups become very small, the spectrum may lose its first order appearance and become <u>SECOND ORDER</u> or <u>HIGHER ORDER.</u> When this occurs, simple, analyzable multiplets are no longer observed.

In aromatic multiplets, several different coupling constants may be observed (J_{ortho}, J_{meta}, J_{para}) which may produce more complex (but still analyzable) multiplets. In cases such as this, the (N + 1) rule is no longer valid but an expanded version of this relationship can be used.

Because of the large size of polymer molecules, a certain amount of band broadening occurs. This loss in resolution is due to the fact that the chemical shifts of the proton groups of each polymer unit are not exactly the same and, as a result, only the most obvious multiplets can be recognized.

In well-resolved NMR spectra, spin-spin coupling provides at least as much, if not more, information than the chemical shifts. In polymer NMR analysis, this is not the case and coupling is generally less important than shift positions.

V.   POLYMER NMR: In the analysis of polymer NMR spectra the basic principles of the method are still employed, namely:

   a.   The chemical shift of the various bands.
   b.   The relative intensities of these bands.
   c.   The overall appearance of the bands.
   d.   To a lesser extent, the multiplicities of the bands.

It should be emphasized here that, save for a loss in resolution, the analysis of polymer NMR spectra is no different from that of other types of spectra. Admittedly, there are some difficulties which are unique to polymers. The main sampling problem is that of low solubility and/or high viscosity resulting in weak or poorly resolved spectra. The use of dilute solutions and/or running the spectrum at higher temperatures does much to alleviate these problems. On the other hand, once the spectroscopist is aware that he is dealing with a polymer, the structural possibilities become sharply limited and analysis is somewhat simplified. The following table has been found helpful in identifying the structural "fragments" which are found in many common, commercial polymers. Being able to recognize the appearance and chemical shifts of these "fragments", will enable the analyst to make a rapid, qualitative identification of many of the less exotic polymer types.

VI.   PREPARING POLYMER SOLUTIONS: The analysis of polymers and resins by NMR requires the preparation of a non-viscous, 20% solution of the sample in a suitable solvent. Although a great deal of solubility information is available for most types of polymers, the solvents suggested are usually not appropriate for preparing NMR solutions. Solvents such as dioxane, cyclohexane, petroleum ether, xylene, tetrahydrofuran, etc., when available in deuterated form, are often quite expensive. During the preparation of the spectra for this volume, the following solvents were employed: carbon tetrachloride, deuterochloroform, acetone-d6, orthodichlorobenzene, deuterium oxide, dimethylsulfoxide-d6, a mixture of deuterochloroform and dimethylsulfoxide-d6 (Polysol-d), trifluoroacetic acid and formic acid.

Trifluoroacetic acid (TFA) was found to be an effective substitute for formic acid in the preparation of solutions of many polyamides and polyesters. Caution in the use of TFA should be employed since decomposition of the sample was observed to occur during its use in the solution of polyoxymethylenes and certain sulfide rubbers. Orthodichlorobenzene used in conjunction with heat was found to be an effective solvent for many of the high molecular weight polyethylene samples.

Two important aids in the preparation of NMR sample solutions are heat and vibration (agitation).

The use of an oil bath and/or a controlled temperature NMR probe will be almost mandatory for many of the more difficultly soluble samples.

Vibrators such as the Cole-Parmer "Super Mixer" can be extremely helpful in dissolving certain slowly soluble samples. When the sample and solvent are placed in a capped vial and gently vibrated for several hours (perhaps overnight), sufficiently concentrated solutions can often be obtained from materials which otherwise appear to be completely insoluble.

SUBSTITUENT (X)	X–CH$_3$	X–CH$_2$–R	X–CH–R$_2$	X–CH$_2$CH$_2$–X	X–CH$_2$–X
R$_3$–Si–	0.0	0.5	- - - -	- - - -	- - - -
R–	0.9	1.3	1.5	1.3	1.3
R–CH=CH–	1.6	1.9	2.2	2.2	2.7
R$_2$N–C(=O)–	2.0	2.2	2.4	2.5	3.5
R–O–C(=O)–	2.1	2.2	2.5	2.6	3.2
N≡C–	2.2	2.4	2.9	- - - -	- - - -
R–S–	2.2	2.5	3.0	2.6	4.0
∅–	2.3	2.9	2.9	2.8	3.8
R–C(=O)–NH–	2.8	3.2	3.8	3.4	4.6
∅–NH–	2.9	3.1	3.6	- - - -	- - - -
Cl–	3.0	3.6	4.0	3.7	5.3
R–O–	3.3	3.4	3.6	3.6	4.7
HO–	3.4	3.5	3.9	3.7	- - - -
Cl–SO$_2$–	3.6	3.8	4.1	- - - -	- - - -
R–C(=O)–O–	3.7	4.2	5.1	4.3	- - - -
∅–O–	3.8	4.0	4.6	4.3	- - - -
F–	4.3	4.4	4.8	- - - -	- - - -
NO$_2$–	4.3	4.4	4.5	- - - -	- - - -

FIGURE I	ADDITION POLYMERS
POLYETHYLENE	$CH_2 = CH_2 \longrightarrow -CH_2CH_2(CH_2CH_2)_nCH_2CH_2-$
POLYVINYL CHLORIDE	$CH_2 = \underset{\underset{Cl}{\vert}}{CH} \longrightarrow -CH_2\underset{\underset{Cl}{\vert}}{CH}\left(CH_2\underset{\underset{Cl}{\vert}}{CH}\right)_n CH_2\underset{\underset{Cl}{\vert}}{CH}-$
POLYMETHYL METHACRYLATE	$CH_2 = \underset{\underset{COOCH_3}{\vert}}{\overset{\overset{CH_3}{\vert}}{C}} \longrightarrow -CH_2\underset{\underset{COOCH_3}{\vert}}{\overset{\overset{CH_3}{\vert}}{C}}-\left(CH_2-\underset{\underset{COOCH_3}{\vert}}{\overset{\overset{CH_3}{\vert}}{C}}\right)_n-CH_2\underset{\underset{COOCH_3}{\vert}}{\overset{\overset{CH_3}{\vert}}{C}}-$
POLYBUTADIENE	$CH_2 = CH - CH = CH_2 \longrightarrow \left(CH_2 - CH = CH - CH_2 \right)_n$

POLYMER CHEMISTRY AND DEFINITIONS

I.   POLYMER CLASSIFICATION AND STRUCTURE:

Although polymers may be classified in many ways, the method of their preparation is one of the most convenient. Some polymers are formed by the addition of one unsaturated unit to another, resulting in the loss of the multiple bonds, or in the case of dienes, in the loss of one of the double bonds. Polymers of this type are classified as ADDITION POLYMERS. The structures of typical addition polymers and the monomers from which they are formed are illustrated in Figure I.

Other macromolecules are synthesized by condensing their monomers to form repeating functional groups (e.g. esters, amides, ethers, etc.) interspersed by alkyl chains or aromatic rings or combinations of both. These condensations are characterized frequently, though not always, by the loss of some by-product (e.g. water, alcohol, etc.). Polymers of this type are described as CONDENSATION POLYMERS. The methods of formation of these polymers are far more varied than those of addition polymers as evidenced by the examples shown in Figure II.

II.   COPOLYMERS

Polymers which are composed of a single repeating unit (HOMOPOLYMER) may not always possess a desired property. Often, by combining two or more monomers into a single polymeric structure, a blending of properties is achieved which results in a product with the desirable feature. Structures of this type are classified as COPOLYMERS. Copolymers may be synthesized by several methods leading to structures of great variety as shown in Figure III:

RANDOM COPOLYMERS: When two or more monomers are mixed together and polymerized simultaneously, they will frequently join together in a random fashion producing a polymer in which there is no predictable sequence of monomeric units. Such polymers lack crystallinity and are often elastic in their properties.

ALTERNATING COPOLYMERS: Certain monomers have a tendency to add to other types of monomers more readily than they do to themselves. When combinations of these compatible monomers are polymerized together, the resulting polymer will consist mainly of regularly alternating monomeric units.

BLOCK COPOLYMERS: A homopolymer may be formed in such a way that one or both of its terminal groups remains chemically active. These active end-groups can then act as initiators for the polymerization of a different monomer at either or both ends of the original chain. The resulting copolymer consists of "blocks" of one homopolymer linked alternately with other homopolymer groups.

9

FIGURE II

CONDENSATION POLYMERS

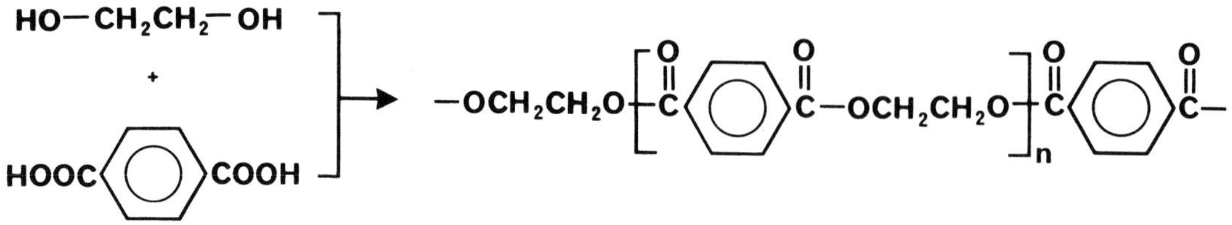

POLYESTER

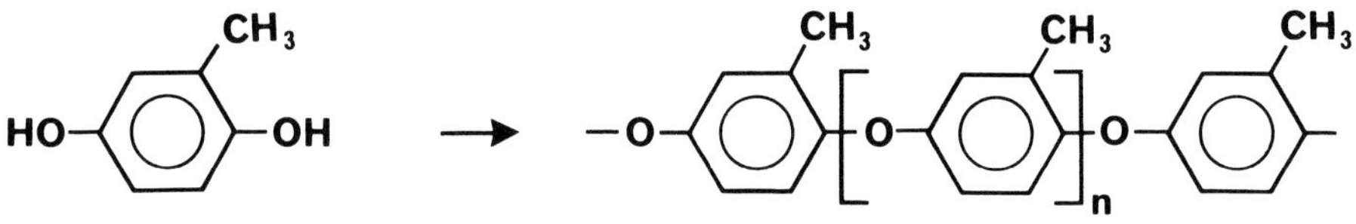

POLYPHENYLENE OXIDE

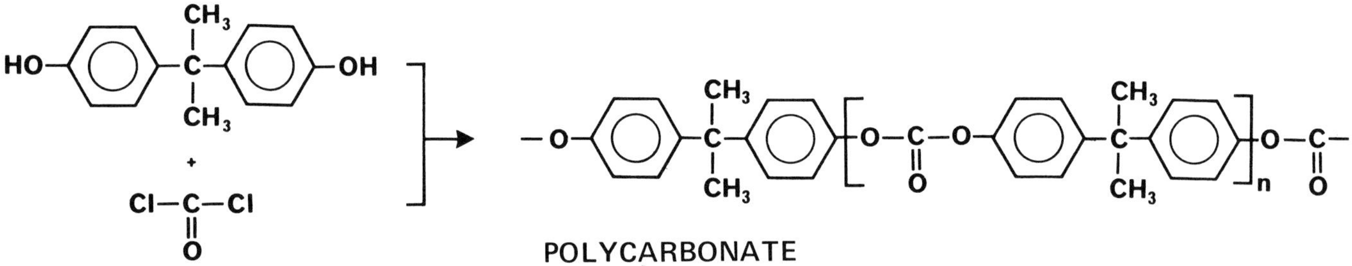

POLYCARBONATE

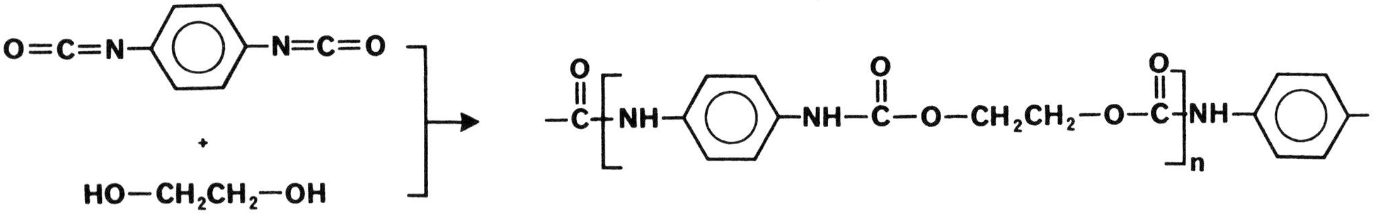

POLYURETHANE

COPOLYMERS **(A,B = Monomer Units)**	<div align="right">FIGURE III</div>
<u>RANDOM</u>	–A–A–B–A–B–B–B–A–B–A–B–A–A–
<u>ALTERNATING</u>	–A–B–A–B–A–B–A–B–A–B–A–B–
<u>BLOCK</u>	–A–A–A–A–B–B–B–A–A–A–A–B–B–B–
<u>GRAFT</u>	–A–A–A–A–A–A–A–A–A–A–A–A– B B B B B B
<u>CROSSLINKED</u>	–A–A–A–A–A–A–A–A–A–A–A–A– B B –A–A–A–A–A–A–A–A–A–A–A–A– B B –A–A–A–A–A–A–A–A–A–A–A–A–

GRAFT COPOLYMERS: These polymers are usually formed by one of two techniques. The first involves the formation of a polymer which contains in its chain other functional groups which can be activated to catalyze the polymerization of a new monomer. The new monomer then adds to the main polymer chain to form pendant side chains which greatly alter the properties of the original polymer. The second method requires the generation of new, active polymerization sites along the chain by use of a catalyst. These active sites then act as focal points for the addition of other polymer chains which are pendant to the main chain. The structure of such a polymer is depicted in Figure III.

CROSSLINKED POLYMERS: There are several techniques available which permit the linking together of many polymer chains to form a giant two or three dimensional matrix. These methods result in a polymer which cannot be readily melted, which is virtually insoluble and is highly resistant to chemical attack. Such a matrix polymer is described as being CROSSLINKED. Crosslinking may be accomplished in several ways. One method is to use a crosslinking agent to tie previously formed chains together by reaction with their pendant functional groups. Another approach is to incorporate a crosslinking agent into the original polymerization mixture. Crosslinking agents are always bi- and trifunctional.

The copolymer types described above are illustrated in Figure III.

III. POLYMER PROPERTIES

The number of monomeric units which make up a polymer chain is termed the DEGREE OF POLYMERIZA-TION. In most polymerization reactions, the degree of polymerization is not uniform for all polymer molecules with a resultant spread in the molecular weights of the various chains. Because of this, when we speak of POLYMER MOLECULAR WEIGHT we are really talking about an average weight. There are two ways of expressing molecular weight; NUMBER-AVERAGE M.W. (M_n) and WEIGHT-AVERAGE M.W. (M_w).

$$(M_n) = \frac{\text{total wt. of all molecules}}{\text{total number of all molecules}}$$

$$(M_w) = \frac{\text{sum (weight of all molecules of a particular M.W. x the M.W. of that type)}}{\text{total weight of all molecules}}$$

Number average M.W. places emphasis on the number of units per given weight of polymer. It can be determined from osmotic pressure and viscosity measurements. Weight average M.W. takes into consideration the total weight of a given number of polymer molecules. It is measured using light scattering, sedimentation studies and ultracentrifugation studies. A weight average M.W. usually gives larger emphasis to the high M.W. fractions since these have a disproportionate effect on light scattering and sedimentation rates.

12

Another important parameter is the ratio (M_w/M_n) since this provides a measure of the uniformity of polymer molecular weight. When the ratio is 1.0, the polymer is said to be <u>MONODISPERSE</u>, meaning that within close limits, all of the chains are of equal length. Monodisperse polymer samples are usually obtainable only through fractionation of the polymer (separating the polymer into narrow M.W. fractions) or in certain cases by using anionic polymerization techniques.

Even so-called "crystalline" polymers contain only sites of crystallinity interspersed with amorphous areas. The polymer chains are randomly coiled and are hard and glassy below a certain temperature. When the polymer is heated, the molecular motions within the chains gradually increase, the chains uncoil, and at a temperature unique for each polymer, the hard solid becomes soft and elastic. The temperature at which this occurs is called the <u>GLASS TRANSITION TEMPERATURE</u> and it is one of the most important physical parameters which is measured. The actual melting of the polymer (the point at which it becomes fluid) may be several hundred degrees higher than the transition temperature. Some polymers have no melting point but begin to degrade when heated above a certain temperature. The fact that many commercial polymers have glass transition points below room temperature is why they are, in fact, "plastic."

In a polymer which is AMORPHOUS, there are no areas of crystallinity <u>(CRYSTALLITES)</u>, as a result the polymer has a low transition temperature and can undergo plastic flow. An <u>UNORIENTED CRYSTALLINE</u> polymer possesses areas of crystallinity which are randomly oriented with respect to each other. When these polymers are stretched along one axis (usually at a temperature above the transition temperature), the crystal- lites become oriented along this axis and the resultant polymer exhibits a much higher tensile strength. Some polymers remain oriented only as long as they are kept under tension and when the tension is removed, they relax back to their unoriented state. Polymers of this type are called <u>ELASTOMERS.</u> The incorporation of crosslinks into a polymer will decrease plastic flow and limited crosslinking enhances elastomeric properties.

When the extent of crosslinking becomes so great in a polymer that it produces a rigid matrix, the result is a <u>THERMOSETTING</u> plastic, since the final crosslinks are established by heating the partially polymerized material until it sets up to a nonmeltable solid. A <u>THERMOPLASTIC</u> resin is one which can be melted and cast repeatedly. Polymers, regardless of their origin, are usually described as fitting one or more of these categories (Thermoplastic, Thermosetting, Elastomeric).

IV. <u>POLYMERIZATION</u>

A chemical which, when added to monomer or a mixture of monomers, starts the polymerization reaction is called a <u>CATALYST</u> or an <u>INITIATOR.</u> The catalyst may be one of three basic types:

(1) A <u>FREE RADICAL</u> initiator

(2) An <u>ANIONIC</u> catalyst

(3) A <u>CATIONIC</u> catalyst

A free radical catalyst is one which undergoes facile homolytic cleavage to form free radicals. These free radicals can then form adducts with a molecule of monomer to produce a monomeric free radical. More monomer can add to this active monomer always resulting in a terminal free radical. Through several processes, the reaction can be finally quenched and an inactive polymer chain is the ultimate result. These catalysts are usually organic peroxides or azo-compounds as shown in Figure IV.

FIGURE IV

FREE RADICAL INITIATORS

BENZOYL PEROXIDE	DI–t–BUTYL PEROXIDE	AZOBISISOBUTYRONITRILE

INHIBITORS may be added to some polymerizations to slow down the rate of polymerization or eventually to terminate it altogether. Quinones are frequently used for this purpose.

Among the anionic catalysts are strong bases such as potassium hydroxide, butyl lithium, alkali metal amides and Grignard reagents. An interesting class of anionic initiators are the Ziegler-Natta catalysts which are titanium chloride-trialkyl aluminum complexes. These catalysts lead to the formation of stereoregular polymers with enhanced physical properties. The introduction of these catalysts into the production of commercial polymers has been one of the biggest scientific breakthroughs in recent years.

Cationic initiators are the least important compounds from a commercial point of view. Strong acids are used, however, to make the urea formaldehyde, urea-melamine and urea-phenol polymers. Other catalysts, which find specific application to the polymerization of alkenes, are aluminum chloride and rhodium chloride. Boron trifluoride has also been utilized for selective polymerizations.

Polymerization is accomplished under a variety of conditions which define the phase in which the reaction procedes. BULK POLYMERIZATION eschews the use of solvent and the resultant polymer melt is pelletized or cast directly. SOLVENT POLYMERIZATION allows the polymerization to occur in a diluted state and results in a viscous polymer solution. In SUSPENSION POLYMERIZATION, the monomer is suspended in globules in an aqueous phase by means of some agent and polymerization is initiated. Each globule acts as a small bulk reaction and solid polymer beads result. EMULSION POLYMERIZATION employs the use of a surfactant and polymerization is initiated within the micelle made up of monomer and emulsifying agent. The water soluble catalyst radical diffuses into the micelle along with fresh monomer to keep the polymerization going. The polymer which results is in the form of beads far smaller (on the order of several hundredths of a micron) than those achieved in bulk polymerization and the polymer remains as an emulsion.

14

FIGURE V

V (a)   ISOTACTIC

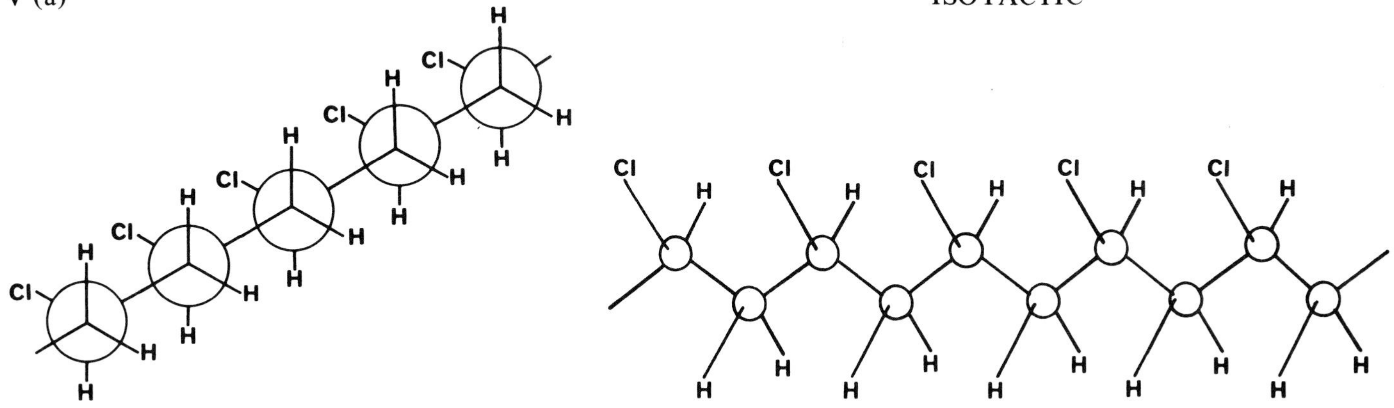

V (b)   SYNDIOTACTIC

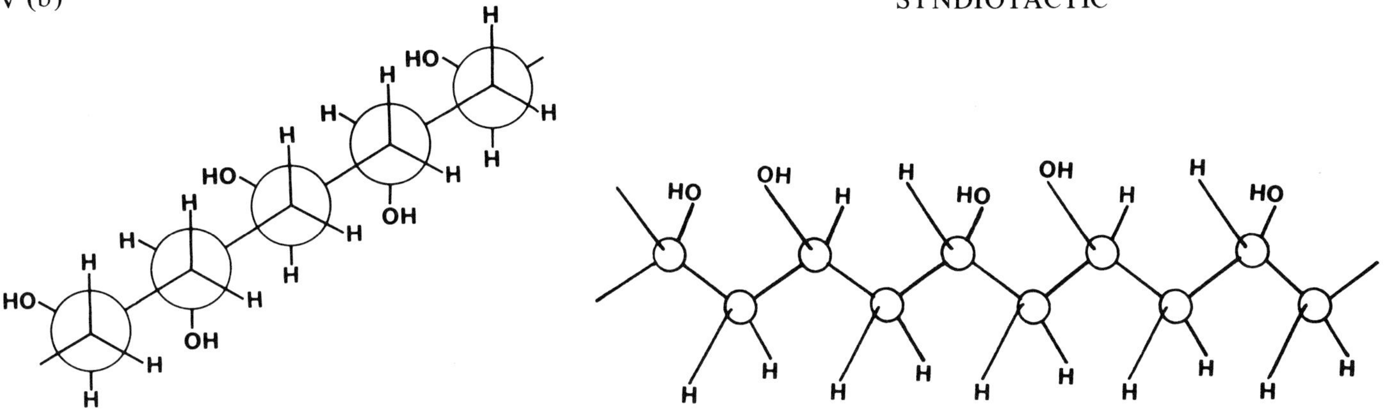

V (c)   ATACTIC

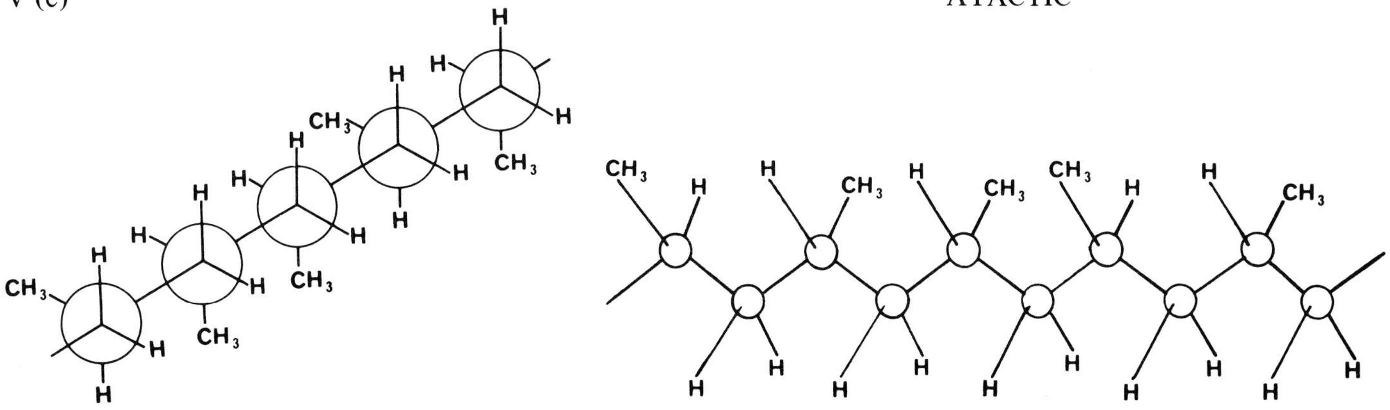

15

V. POLYMER-CHAIN CONFIGURATION

When substituted vinyl compounds are polymerized, the carbon atoms in the polymer chain which bear the substituents, become PSEUDO-ASYMMETRIC. That is, they are asymmetric but not extensively so since two of the groups leading to this asymmetry are essentially identical chains differing only in the length of the chain. However, the introduction of these repetitive sites of "asymmetry" along the chain allows for differences in the sequence configuration or STEREOREGULARITY of the polymer. The type and extent of this stereoregularity in the polymer is called TACTICITY.

If all of the monomeric units possess the same enantiomorphic configuration, then the polymer is described as ISOTACTIC. As illustrated in Fig. V(a), all of the R-substituents appear on the same side of the chain. When the enantiomorphic configuration along the chain alternates regularly, the polymer is described as being SYNDIOTACTIC. In Fig. V(b), the R-groups are seen to alternate regularly from one side of the chain to the other along the polymer backbone. When the substituents (and hence the configurations) at each monomer are randomly placed, the resultant polymer is described as ATACTIC. In polymers which can exhibit tacticity, the extent of the stereoregularity determines the crystallinity and the physical properties of the polymer. The presence or absence of tacticity, and the type of tacticity, is controlled by the catalyst employed in the polymerization reaction.

Common polymers which can be prepared in specific configurations include poly(-olefins), polystyrene, polymethyl methacrylate, poly(1,2-butadiene) and polyvinyl isobutyl ether.

All three stereochemical varieties of polymethylmethacrylate are known. The free radical initiated polymer is atactic and amorphous. Various anionic catalysts can be used to form either crystalline isotactic or crystalline syndiotactic products. Of the three, the isotactic polymethylmethacrylate has the highest glass transition temperature and the syndiotactic the lowest. In general, free radical catalysis usually leads to atactic polymers while anionic or Ziegler-Natta catalysts produce stereoregular chains.

Obviously, many polymers are neither completely stereoregular nor atactic. The description of the stereoregularity of short chain sequences (e.g. 2-5 units) of a polymer is termed MICROTACTICITY. This description becomes somewhat involved because here we are attempting to describe the relative configurations of short chain sequences.

In using proton NMR for polymer analysis, we find for many polymers that we can distinguish the Diad and Triad structures for these materials. In Fig. VI(a) we see a diad sequence (two monomer units) in which protons H_a and H_b are non-equivalent due to their spatial relationship to the X-substituents on the adjacent carbon atoms. Since the two asymmetric carbon atoms on either side of the methylene group have the same configuration, we designate this diad meso or dd or 11. In Fig. VI(b), both methylene protons H_a are equivalent and this type of diad is termed racemic or d1.

16

```
   H   H   Hb  H                          H   H   Ha  X
   |   |   |   |                          |   |   |   |
 - C - C - C - C -                      - C - C - C - C -
   |   |   |   |                          |   |   |   |
   H   X   Ha  X                          H   X   Ha  H
```

a meso or dd or 11 DIAD   a racemic of dl DIAD

Fig. VI(a)   Fig. VI(b)

In the analysis of <u>Triads</u> the proton attached to the asymmetric carbon atom becomes the focal point and we view its position in relation to the functional groups (X) on adjacent monomer units. In Fig. VII(a), H_a is located opposite to both X-groups. This is described as an <u>Isotactic Triad.</u> In Fig. VII(b), H_b is on the same side as both substituents; a <u>Syndiotactic Triad.</u> Fig. VII(c) shows H_c on the same side as one X-group and on the opposite side to the other; a <u>Heterotactic Triad.</u>

```
   H   H   Ha  H   H   H                  H   H   X   H   H   H
   |   |   |   |   |   |                  |   |   |   |   |   |
 - C - C - C - C - C - C -              - C - C - C - C - C - C -
   |   |   |   |   |   |                  |   |   |   |   |   |
   X   H   X   H   X   H                  X   H   Hb  H   X   H
```

Fig. VII(a)   Fig. VII(b)

```
   H   H   X   H   X   H
   |   |   |   |   |   |
 - C - C - C - C - C - C -
   |   |   |   |   |   |
   X   H   Hc  H   H   H
```

Fig. VII(c)

Another way in which polymer chain configurations can differ is by exhibiting geometric isomerism. If the various 1,3-diene monomers are polymerized under proper conditions, they form mainly 1,4-adducts with residual double bonds along the backbone of the chain. Since each double bond is at least doubly substituted, the substituents can exist either cis- or trans- to each other. An all cis-polymer will exhibit very different properties from an all trans-polymer. Even nature takes cognizance of this fact as exemplified by natural rubber which is cis-polyisoprene and gutta-percha which is trans-polyisoprene.

Suffice it to say, that the use of NMR and other physical methods can identify such fine detail in polymer structure and be of great importance in theoretical considerations of polymer properties.

Commercial Designations

The methods used by the polymer industry to name their commercial products is no more consistent than that used by the manufacturers of other chemical products.

Often, a well known trademark name will be applied to a wide variety of different polymers and resins with only the addition of numbers and/or letter suffixes to indicate a specific product. For example, the name "Bakelite" (Union Carbide Corporation) has been used to describe a whole spectrum of plastics and resins which include vinyl acetates, vinyl ethers, vinyl chlorides, polyethylenes and phenolic resins. In other instances, a trademark name may be restricted to just one basic type of polymer or product as with the name "Lucite" (E. I. du Pont de Nemours & Co.) which is used almost exclusively for various esters of methacrylic acid and those final products which contain these polymers.

In regard to the generic names of textile fibers, the Federal Trade Commission has specifically defined many of the names which are applied to synthetic fabrics. The term "acetate fiber" is limited to those fibers which are made from cellulose acetate. In order for the term "triacetate" to be used, the polymer must be at least 92% acetylated. The term "modacrylic" can only be applied to those fibers which contain at least 35% but less than 85% by weight of acrylonitrile units.

Another naming method which may be of some aid in the characterization of polymers is that in which the abbreviation of the polymer name is used as the manufacturers designation. The commercial products listed below illustrate this procedure.

CMC 120	a carboxymethyl cellulose	Hercules, Inc.
DAPON 35	a diallyl phthalate resin	FMC Corporation
EPR 404	an ethylene-propylene rubber	Enjay Chemical Co.
PE 204C	a polyethylene	Rexall Chemical Co.
PEG 1000	a polyethylene glycol	Allied Chemical Co.
PVP K-30	a polyvinyl pyrrolidone	GAF Corporation

A listing of monomer and polymer abbreviations which are in common use begins on the next page.

MONOMER AND POLYMER ABBREVIATIONS

ABS	Acrylonitrile-butadiene-styrene
ASA	Acrylic ester-styrene-acrylonitrile
BACN	Butadiene-acrylonitrile
BPA	Bisphenol-A (2,2-diphenylol propane)
BR	Butadiene rubber
CA	Cellulose acetate
CAB	Cellulose acetate butyrate
CFE	Poly(chlorofluoroethylene)
CMC	Carboxymethyl cellulose
CMHEC	Carboxymethyl hydroxyethyl cellulose
CR	Chloroprene rubber
DADI	Dianisidine diisocyanate
DAP	Diallyl phthalate
DMT	Dimethyl terephthalate
DMU	Dimethylol urea
DVB	Divinyl benzene
EG	Ethylene glycol
EPDM	Ethylene-propylene-diene monomer
EPM	Ethylene-propylene-monomer
EPR	Ethylene-propylene rubber
EPT	Ethylene-propylene terpolymer
EVA	Ethylene-vinyl acetate
EVE	Ethyl vinyl ether
FBA	Perfluorobutyl acrylate
FEP	Fluorinated ethylene-propylene
FRP	Fiber reinforced plastic
HDI	Hexamethylene diisocyanate
HEMA	Hydroxymethyl methacrylate
HET	Hexachloroendomethylenetetrahydrophthalic acid (anhydride)
HPA	Hydroxypropyl acrylate

IVE	Isobutyl vinyl ether
LPE	Linear polyethylene
MBS	Methyl methacrylate-butadiene-styrene
MDI	Methylene di-p-phenylene isocyanate
MF	Melamine-formaldehyde
MMU	Monomethylol urea
NBR	Nitrile-butadiene rubber
NC	Nitrocellulose
NDI	1,5-Naphthalene diisocyanate
NR	Nitrile rubber
NSR	Nitrile-silicone rubber
PA	Polyamide; polyacrylate; polyallomer
PAN	Polyacrylonitrile
PAT	Polyaminotriazole
PBAA	Polybutadiene-acrylic acid
PBI	Polybenzimidazole
PC	Polycarbonate
PCP	Polychloroprene
PCTFE	Poly(chlorotrifluoroethylene)
PDI	Phenylene diisocyanate
PE	Polyethylene; pentaerythritol
PEG	Polyethylene glycol
PET	Polyethylene terephthalate
PETP	Polyethylene terephthalate
PF	Phenol-formaldehyde
PG	Polypropylene glycol
PIB	Polyisobutylene; polyisobutene
PMA	Polymethyl acrylate
PMMA	Polymethyl methacrylate
PMP	Polymethyl pentene
POE	Polyoxyethylene
POEOP	Polyoxyethylene-oxypropylene
PP	Polypropylene

PPO	Polypropylene oxide; polyphenylene oxide (46)
PPX	Poly-p-xylylene
PS	Polystyrene
PTFE	Poly(tetrafluoroethylene)
PTSA	p-Toluenesulfonamide
PVA	Polyvinyl acetate; polyvinyl alcohol
PVAc	Polyvinyl acetate
PVAC	Polyvinyl acetate
PVAl	Polyvinyl alcohol
PVAL	Polyvinyl alcohol
PVB	Polyvinyl butyral
PVC	Polyvinyl chloride
PVdC	Polyvinyl dichloride
PVDC	Polyvinyl dichloride
PVDF	Polyvinyl difluoride
PVE	Polyvinyl ethyl ether
PVF	Polyvinyl fluoride
PVI	Polyvinyl isobutyl ether
PVM	Polyvinyl methyl ether
PVM/MA	Polyvinyl methyl ether-maleic anhydride
PVOH	Polyvinyl alcohol
PVP	Polyvinyl pyrrolidone
RDGE	Resorcinol diglycidyl ether
SA	Styrene-acrylonitrile
SAN	Styrene-acrylonitrile
SBR	Styrene-butadiene rubber
SCMC	Sodium carboxymethyl cellulose
TA	Terephthalic acid
TDI	Tolylene diisocyanate
TFE	Poly(tetrafluoroethylene)
TODI	Tolidine diisocyanate
TPA	Terephthalic acid
TDQP	Trimethyl dihydroquinoline polymer

UF	Urea formaldehyde
VC	Vinyl chloride; vinylidene chloride
XDI	Xenylene diisocyanate
XLPE	Crosslinked polyethylene

THE HYDROCARBON POLYMERS

The polymers in this section contain only carbon and hydrogen. They have been divided into three sub-categories based upon spectroscopic rather than chemical criteria since each of these subdivisions possess distinctly different NMR parameters.

(1)   Saturated aliphatic hydrocarbons $(-CH_2-CH_2-)_n$

(2)   Olefinic/aliphatic hydrocarbons $(-CH_2-CH=CH-CH_2-)$

(3)   Aromatic/aliphatic hydrocarbons $(-CH_2-CH-)_n$

All of the monomers from which these polymers are prepared contain at least one double bond which is lost during the polymerization process. If two double bonds are present in the monomer two or more different polymers units may be produced. The polymerization of butadiene, for example, can result in three different isomers (1,4-cis-, 1,4-trans-, 1,2-polybutadiene).

Virtually all of the commercial polymers were originally formed using free radical initiators, however, in recent times stereospecific polymerizations using Ziegler-Natta catalysts have become common.

NMR Parameters

The saturated aliphatic polymers are characterized by the appearance of bands within the chemical shift range from 0.8 to 2.0 ppm delta. They may display very simple spectra such as those of the polyethylenes (one single peak) or polyisobutylenes (two single peaks) or may produce quite complex spectra such as those of the poly-1-butenes. The main feature of their spectra and the one which distinguishes them from others in this hydrocarbon class, is the narrow range within which all of their bands occur (about 1.0 ppm).

The olefinic/aliphatic hydrocarbon polymers, which contain residual double bonds, exhibit resonance bands in the chemical shift range from 4.7 to 5.4 ppm in addition to the aliphatic bands at high field (0.8 - 2.0 ppm). Within the olefinic region (4.7 - 5.4 ppm), it is possible to distinguish between terminal olefinic protons (4.7 - 5.2 ppm) and the internal, backbone vinyl protons which resonate at about 5.4 ppm. The saturated proton groups (methylenes and methines) which are adjacent to the double bonds normally resonate slightly below 2.0 ppm.

The spectra of the <u>aromatic/aliphatic hydrocarbon polymers</u> display characteristic bands in the 6.5 to 7.2 ppm region. Styrene homopolymers and block copolymers exhibit two broad, symmetrically shaped peaks at 6.5 and 7.1 ppm. The integration ratio of these bands is 2:3. When the styrene units alternate with another monomer along the chain, then only one aromatic band is usually seen which possesses a chemical shift of about 7.1 ppm.

The <u>solubility characteristics</u> of the hydrocarbon polymers vary greatly depending directly upon the molecular weight of the polymer and the degree of crosslinking which is present. Commercial samples of the saturated aliphatic polymers dissolve with difficulty only in hot aromatic solvents such as ortho-dichlorobenzene. The polymers containing residual olefinic hydrogens and aromatic rings are often readily soluble in chlorinated solvents such as deuterochloroform.

<u>Nomenclature and Commercial Products</u>

All of the synthetic hydrocarbon polymers take their common names from the monomers from which they were formed (ethylene – polyethylene). Listed below are some of the common and commercial names and abbreviations which are applied to certain hydrocarbon polymer types. The numbers in parentheses refer to the list of manufacturers on page 275 to whom the trademark names belong.

    <u>Polyethylene</u>
        Alathon (1), Ethafil (2), Ethofoam (3), Ethron (3), Marlex (4), polythene, PE.

    <u>Polypropylene</u>
        Herculon (5), Marlex (4), Olefane (6), polypropene, polymethylethylene, PP.

    <u>Ethylene/propylene copolymers</u>
        Nordel (1), Royalene (10), Vistalon (8), EP, EPM, EPDM, EPR, EPT.

    <u>Polyisobutylene</u>
        Oppanol (7), Vistanex (8), butyl rubber, polybutene, polyisobutene, poly-2-methylpropene, PIB.

    <u>Polymethyl pentene</u>
        TPX (9), polyisobutylethylene, poly-4-methylpentene-1, PMP.

Polybutadiene

Ameripol CB (11), Budene (12), Budium (1), Cis-4 (4), Cisdene (13), Diene (14), Plioflex (12), Synpol (15).

Polyisoprene

Ameripol SN (11), Coral (14), Natsyn (12), poly(2-methyl-1,3-butadiene), natural rubber.

Polystyrene

Dylene (16), Dylite (16), Lustrex (17), Piccolastic (18), Styrofoam (3), Styron (3), polyvinyl-benzene, polyphenylethylene, PS.

Styrene/butadiene copolymers

Ameripol 4502, 4503 (11), Baytown (19), Duradene (14), FR-S (14), Naugapol (10), Phil-prene (4), Plioflex (12), Solprene (4), Synpol (15), SBR.

HYDROCARBON POLYMERS

The low molecular weight of this polyethylene sample is indicated by its ready solubility in deutero-chloroform and the observance of a significant methyl absorbance band which appears as a distorted triplet at 0.89 ppm. The methylene groups in the chain possess very similar chemical shifts and overlap to form a broad single peak at about 1.27 ppm. Based upon the observed integration values, the ratio of methylene groups to methyl groups is about 11.6 to 1 which would indicate a molecular weight of approximately 350 for this material.

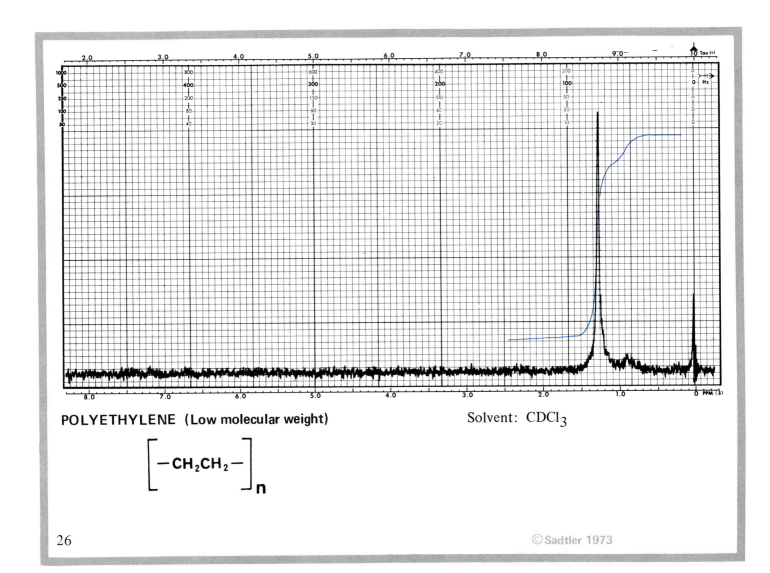

POLYETHYLENE (Low molecular weight)

Solvent: $CDCl_3$

$$\left[-CH_2CH_2- \right]_n$$

The characteristic absorbance band of the polyethylenes is an intense sharp single peak near 1.3 ppm which represents the resonance of the chain methylene groups. The methyl groups present in the polymer as terminating groups or on short branches, appear as a broadened triplet near 0.9 ppm. This methyl band is often difficult to observe in the spectra of high density polyethylenes (few branching groups) but becomes significant in the spectra of low density polyethylenes which may contain up to forty branching groups.

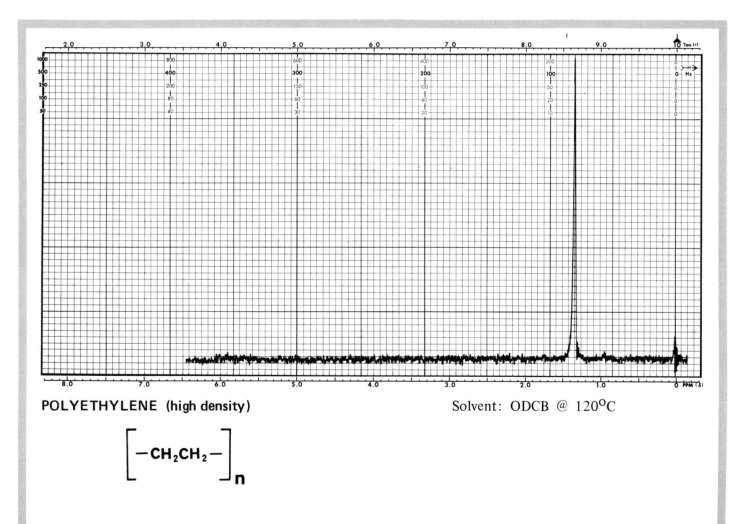

POLYETHYLENE (high density)

Solvent: ODCB @ 120°C

$$\left[-CH_2CH_2-\right]_n$$

HYDROCARBON POLYMERS

The NMR spectra of the polypropylenes display a characteristic "fingerprint" pattern in the chemical shift range from 0.6 to 2.0 ppm. The methyl groups of the polymer resonate as a broadened doublet centered at about 0.84 ppm while the methylene and methine resonances form overlapping multiplets centered at 1.1 and 1.6 ppm respectively. The chemical shift range of the polypropylenes is similar to that of polymethyl pentene (page 30) and certain polybutenes (pages 32, 33, 34).

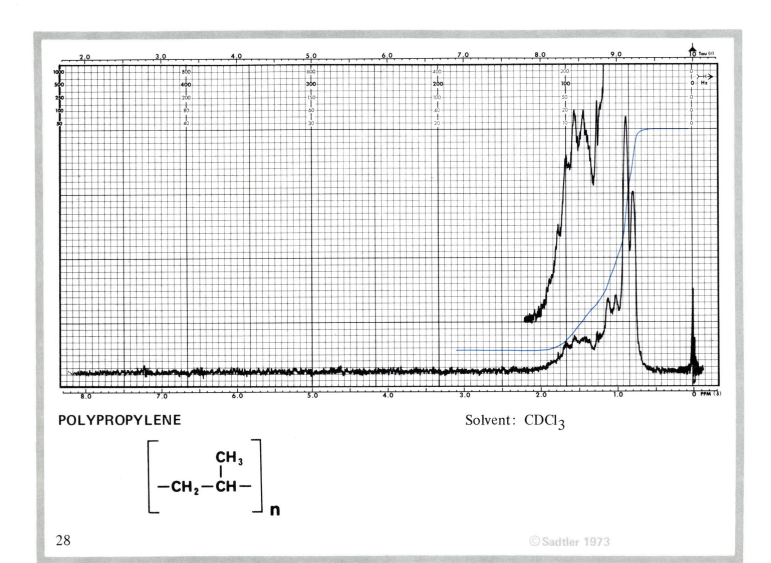

POLYPROPYLENE   Solvent: $CDCl_3$

$$\left[-CH_2-\underset{\underset{CH_3}{|}}{CH}- \right]_n$$

Polyisobutylene produces an NMR spectrogram consisting of two sharp single peaks. The equivalent methyl groups resonate as a six proton singlet at 1.11 ppm while the methylene groups appear as a two proton singlet at 1.41 ppm. There are no bands observed in the spectrogram to indicate the presence of residual unsaturation in the polymer chain nor to suggest the type of terminating groups which are present. The relative simplicity and "cleanness" of this spectrogram are characteristic of polyisobutylene and make this polymer particularly easy to identify.

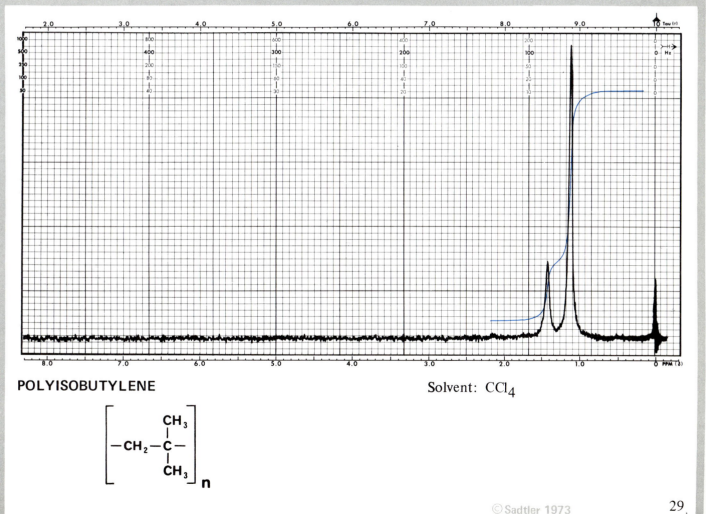

POLYISOBUTYLENE

Solvent: CCl_4

29.

Polymers of 4-methyl-1-pentene produce NMR spectra similar in general appearance to those of poly-propylene. The methylene and methine groups resonate over the chemical shift range from 1 to 2 ppm while the methyl groups appear at 0.95 ppm as an intense doublet. The narrow line width of this doublet may be helpful in distinguishing the NMR spectra of polymethyl pentenes from those of polypropylenes.

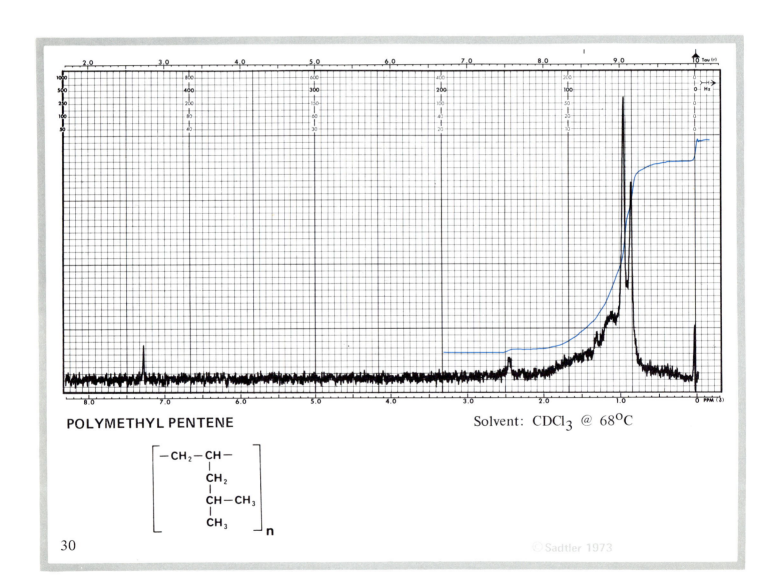

POLYMETHYL PENTENE

Solvent: $CDCl_3$ @ 68°C

$$\left[\begin{array}{l} -CH_2-CH- \\ \qquad\quad | \\ \qquad\quad CH_2 \\ \qquad\quad | \\ \qquad\quad CH-CH_3 \\ \qquad\quad | \\ \qquad\quad CH_3 \end{array} \right]_n$$

The characteristic polyethylene band is observed at 1.26 ppm and the polypropylene methyl groups resonate at slightly higher field as a poorly resolved doublet. Based upon a comparison of the integration values, the ratio of ethylene units to propylene units is approximately 2 to 1. Inclusion of a small amount of a diene into an ethylene-propylene copolymer such as this yields a terpolymer that can be vulcanized with sulfur. The amount of diene present in such terpolymers is usually so small that its absorption bands are not detected.

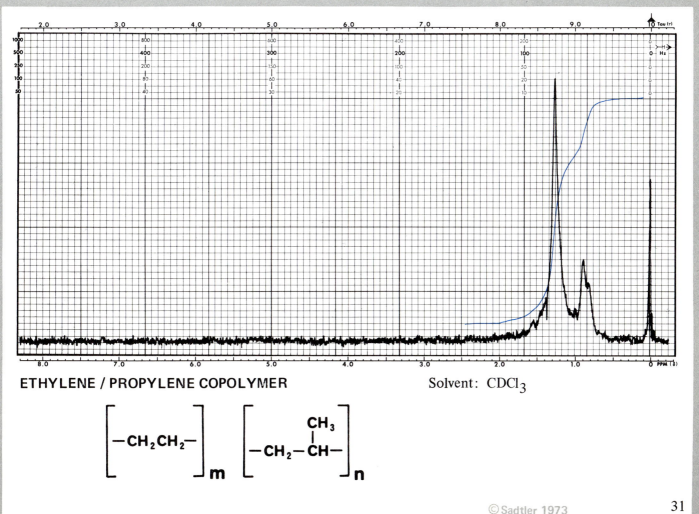

ETHYLENE / PROPYLENE COPOLYMER   Solvent: $CDCl_3$

31

HYDROCARBON POLYMERS

The presence of the characteristic bands at 1.1 and 1.4 ppm indicate that this low molecular weight poly-butene (M.W. about 875) contains a high percentage of isobutylene units. The additional bands which are observed at 0.88, 1.00 and 1.32 ppm probably represent the protons of 1-butene units which are present in the polymer. The absence of absorption bands in the chemical shift range from 5 to 6 ppm indicates that there is no appreciable amount of residual unsaturation present in this sample.

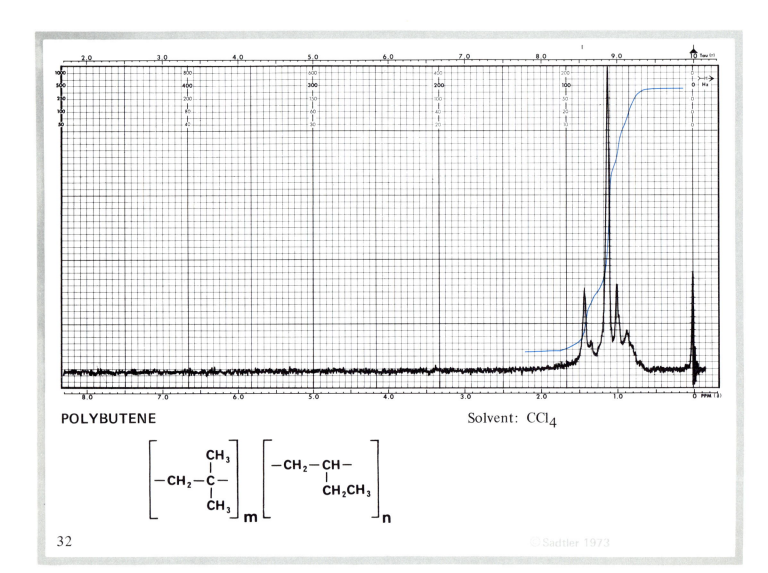

POLYBUTENE

Solvent: CCl$_4$

The appearance of high field bands at 0.85 and 1.00 ppm indicates that this low molecular weight polybutene (M.W. about 400) contains a relatively high percentage of 1-butene units in addition to those of isobutylene. The bands at 1.1 and 1.4 ppm which represent the isobutylene units in this polymer are dwarfed in comparison with the bands arising from the 1-butene units. The polybutenes are used as hot melt adhesives, polymer modifiers and lubricant additives.

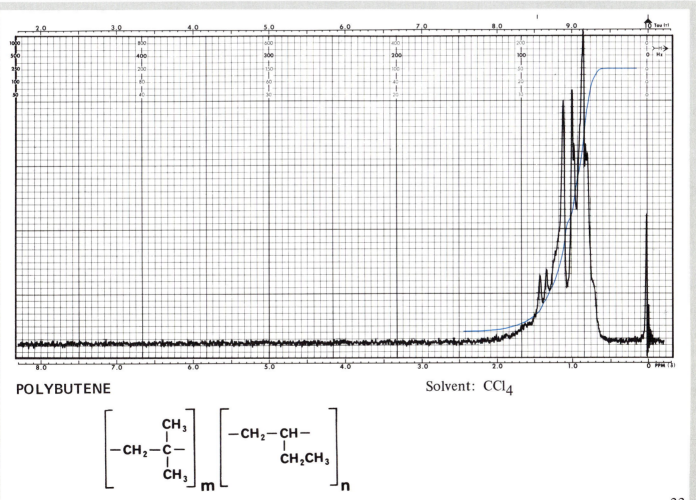

POLYBUTENE

Solvent: CCl_4

$$\left[-CH_2-\underset{\underset{CH_3}{|}}{\overset{\overset{CH_3}{|}}{C}}- \right]_m \left[-CH_2-\underset{\underset{CH_2CH_3}{|}}{CH}- \right]_n$$

33

HYDROCARBON POLYMERS

Although similar in general appearance to the previous spectrum, the appearance of additional bands at 2.0 and 5.2 ppm indicate the presence of carbon-carbon double bonds in the polymer chain. The band at 2.0 probably represents methyl or methylene groups adjacent to the double bond while the band at 5.2 would represent the resonance of the olefinic protons.

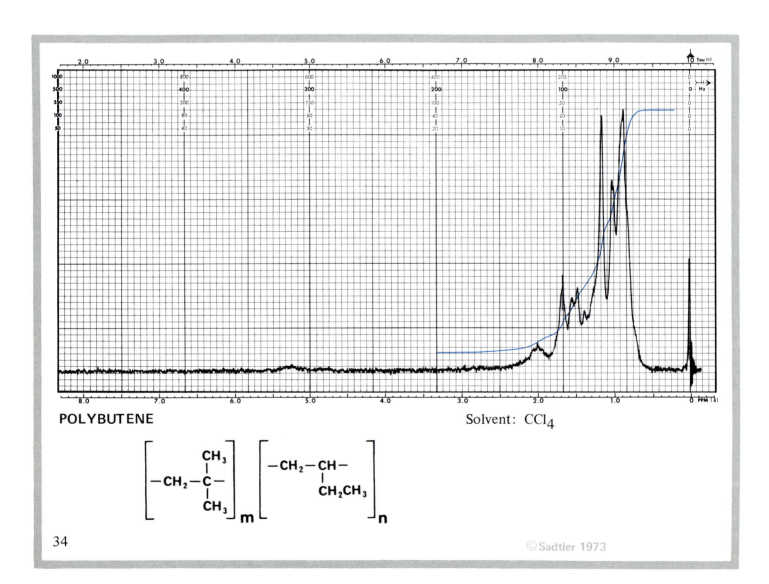

POLYBUTENE

Solvent: CCl₄

$$\left[-CH_2-\overset{\displaystyle CH_3}{\underset{\displaystyle CH_3}{\overset{|}{\underset{|}{C}}}}- \right]_m \left[-CH_2-\overset{\displaystyle CH-}{\underset{\displaystyle CH_2CH_3}{\overset{}{\underset{|}{}}}} \right]_n$$

34

The methylene groups are deshielded by the double bond and resonate at 2.09 ppm as a broadened singlet. The vinyl hydrogens absorb near 5.4 ppm. The chemical shifts and appearance of the bands of the trans isomer are virtually identical to those of the cis isomer when examined on a 60 MHz NMR spectrometer. The very weak band at 5.08 ppm is due to the presence of a small percentage of 1,2-polybutadiene (see page 37).

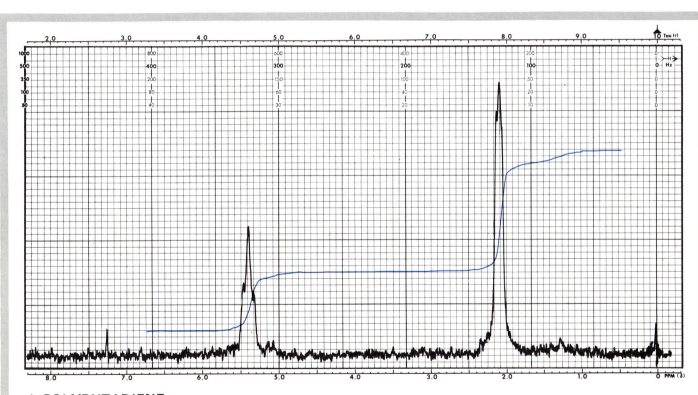

cis-POLYBUTADIENE

This NMR spectrum of trans polybutadiene is practically indistinguishable from that of the cis isomer. Both spectra display a broadened band near 2.02 ppm arising from the methylene groups adjacent to the carbon-carbon double bonds and weaker bands at 5.41 representing the olefinic protons in each unit. Although this spectrum does not display the same type of fine structure as that of the cis isomer, this difference appears to be related more to concentration and viscosity effects than to the geometrical differences between the cis and trans structures.

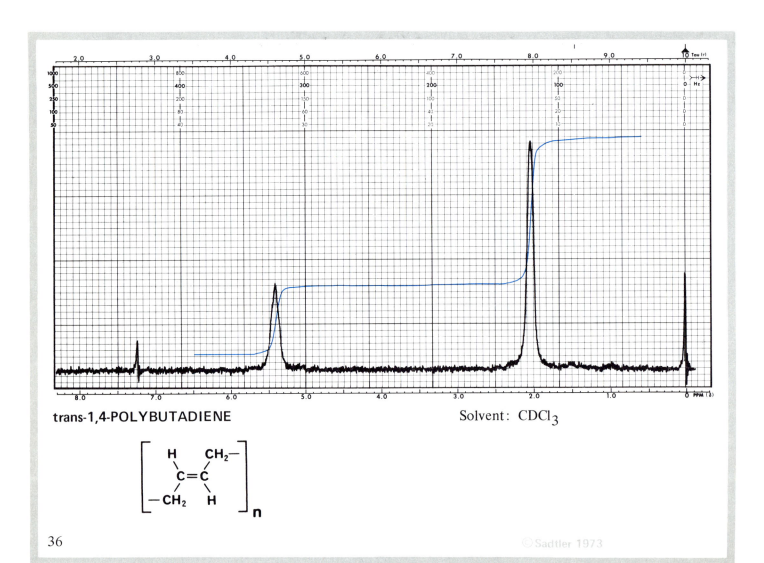

trans-1,4-POLYBUTADIENE   Solvent: CDCl₃

The pendant vinyl groups result in a characteristically complex set of absorbance bands in the chemical shift range from 4.6 to 6.0 ppm. The bands at 2.05 ppm and 5.38 ppm suggest that the sample contains a relatively high percentage of 1,4-polybutadiene units. A comparison of the integration ratios indicates that the composition of the sample is 68% 1,2-polybutadiene and 32% 1,4-polybutadiene (page 35).

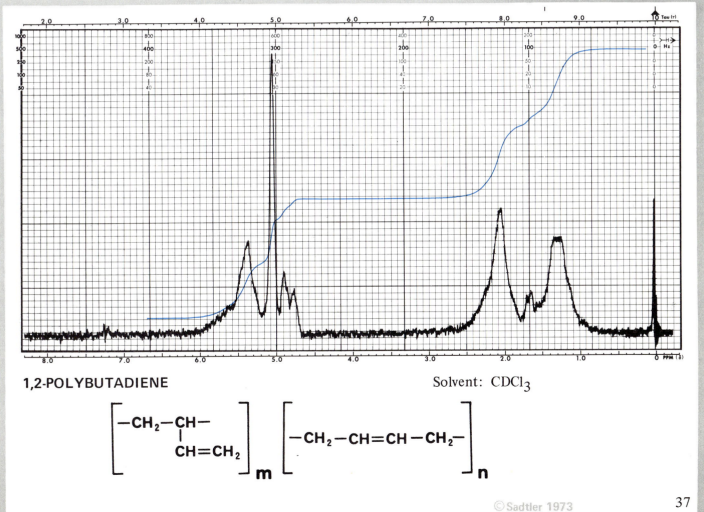

1,2-POLYBUTADIENE

Solvent: $CDCl_3$

$$\left[-CH_2-CH- \atop CH=CH_2 \right]_m \left[-CH_2-CH=CH-CH_2- \right]_n$$

37

HYDROCARBON POLYMERS

Commercial carboxylated butadiene rubbers are composed of 1,4-butadiene (about 70%), 1,2-butadiene (about 28%) and methacrylic acid (about 2%). The strong bands at 2.09 ppm and 5.43 ppm represent the 1,4-butadiene units while the bands in the chemical shift range from 4.6 to 5.2 ppm arise from the 1,2-butadiene units. The only indication of the presence of methacrylic acid units in the spectrum is the observance of several weak peaks in the chemical shift range from 1 to 3 ppm.

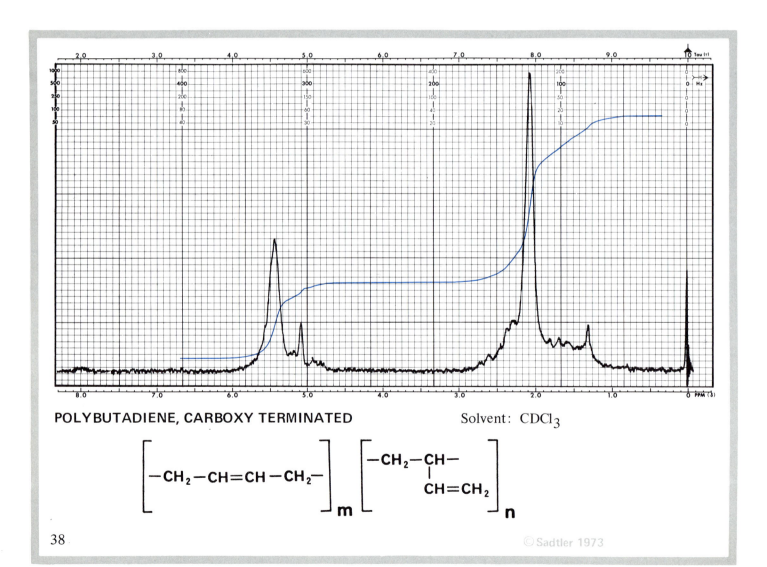

POLYBUTADIENE, CARBOXY TERMINATED   Solvent: $CDCl_3$

$$\left[-CH_2-CH=CH-CH_2- \right]_m \left[\begin{array}{c} -CH_2-CH- \\ | \\ CH=CH_2 \end{array} \right]_n$$

38

Cis-1,4-polyisoprene produces a relatively simple NMR spectrogram consisting of three major absorbance bands. The methyl groups resonate at 1.75 ppm as a sharp single peak. The two different types of methylene groups possess very similar chemical shifts and overlap to produce a broad single peak at about 2.12 ppm while the olefinic protons appear as a weak broadened triplet at about 5.25 ppm.

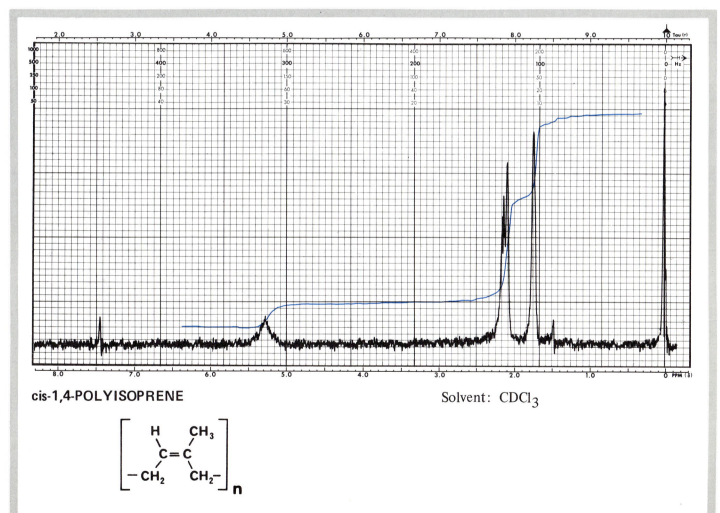

cis-1,4-POLYISOPRENE

Solvent: $CDCl_3$

39

HYDROCARBON POLYMERS

The major absorbance bands of this spectrogram arise from the protons of cis-1,4-polyisoprene which is the primary component of natural rubber. Those bands are observed at 1.72 ppm, 2.15 ppm and 5.18 ppm. The only significant difference between this sample and that of cis-1,4-polyisoprene on page 39 was noted to be the difference in sample solubility. The synthetic material was found to be readily soluble in $CDCl_3$ while the natural product required the use of orthodichlorobenzene and elevated temperatures in order to obtain a reasonable solution concentration.

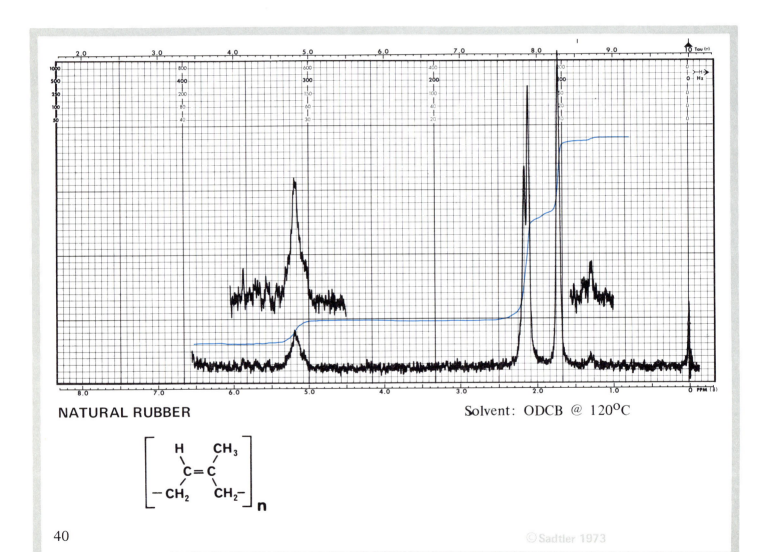

NATURAL RUBBER

Solvent: ODCB @ 120°C

The NMR spectrogram of this low molecular weight polymer of propylene contains absorbance bands arising from both aliphatic and olefinic protons. The polypropylene aliphatic groups appear as a characteristic series of peaks in the chemical shift range from 0.8 to 1.8 ppm. The bands which resonate at 4.7 and 5.2 ppm represent the protons bonded to residual double bonds in the sample. The band at 4.75 ppm probably represents olefinic protons of the type $H_2C=C-CH_3$ and the band at 5.2 ppm may result from the olefinic protons of the fragment $-CH=CH-CH_3$.

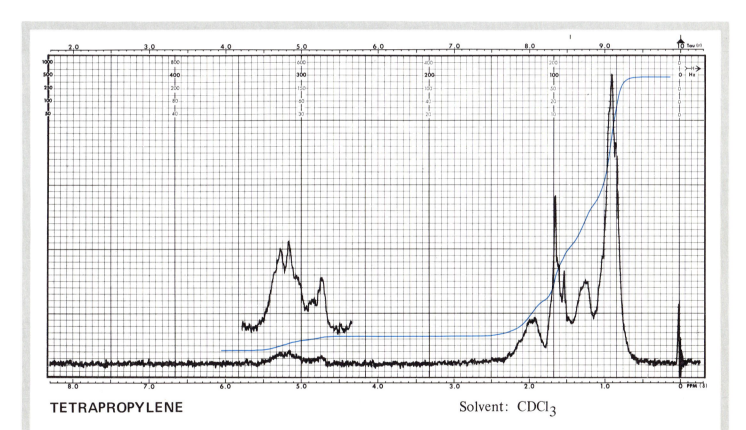

TETRAPROPYLENE

Solvent: $CDCl_3$

COMPLEX STRUCTURE

41

The methylene and methine protons of the polymer chain form a broad overlapping absorbance band in the chemical shift range from 1.1 to 2.4 ppm. The aromatic ring hydrogens are represented by two broad bands at lowest field. The band at 6.6 ppm represents the two ortho protons and the band at 7.06 arises from the meta and para hydrogens. When styrene units alternate positions with another type of unit, as in a copolymer, the ortho hydrogens are not preferentially shielded and a single peak is observed for all five ring hydrogens. They then appear as a single broad band at about 7.06 ppm as in the spectrum of the styrene/butadiene copolymer depicted on page 46.

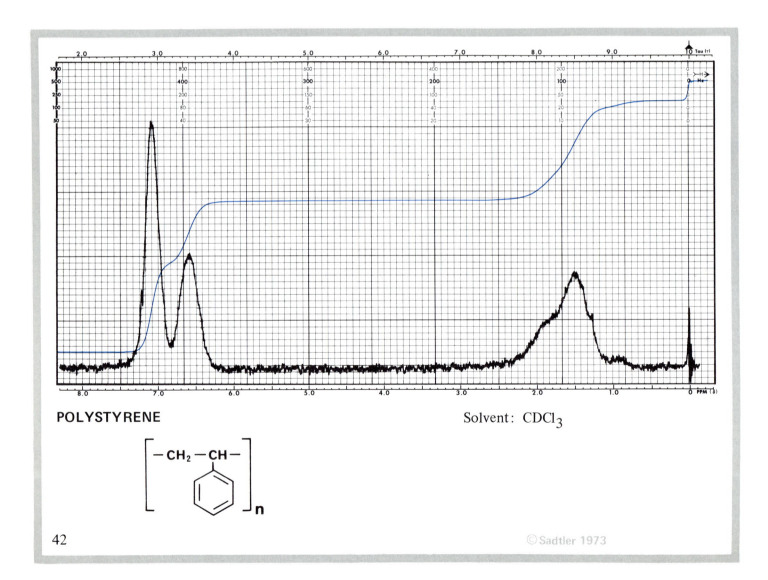

POLYSTYRENE   Solvent: $CDCl_3$

$$\left[-CH_2 - CH - \right]_n$$

Although similar in general appearance to the spectrum on the previous page, this spectrum of stereo-regular isotactic polystyrene displays slight changes in chemical shift. The band representing the ortho aromatic hydrogens resonates at slightly higher field (6.59 vs 6.61 ppm) and the methine absorption band resonates at slightly lower field (2.0 vs 1.88 ppm).

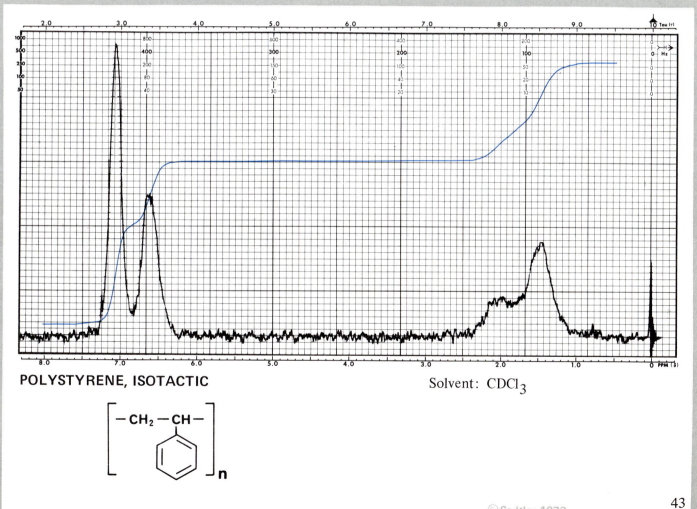

POLYSTYRENE, ISOTACTIC

Solvent: $CDCl_3$

$$\left[-CH_2-CH- \right]_n$$

43

In addition to the typical polystyrene absorption bands at 1.1 to 2.3, 6.5 and 7.01 ppm, the spectrum of this sample contains bands at 0.7 to 1.1 ppm which are quite narrow in line width. The fact that they are more highly resolved than those of the polystyrene bands suggests that these bands may be due to a low molecular weight additive, such as a plasticizer. Since the plasticizer's complete chemical shift range is no greater than from 0.7 to 2.2 ppm, it is probably an unsubstituted, saturated hydrocarbon material. The complexity of the methyl absorbance band near 0.9 ppm indicates that it probably contains aliphatic branching groups.

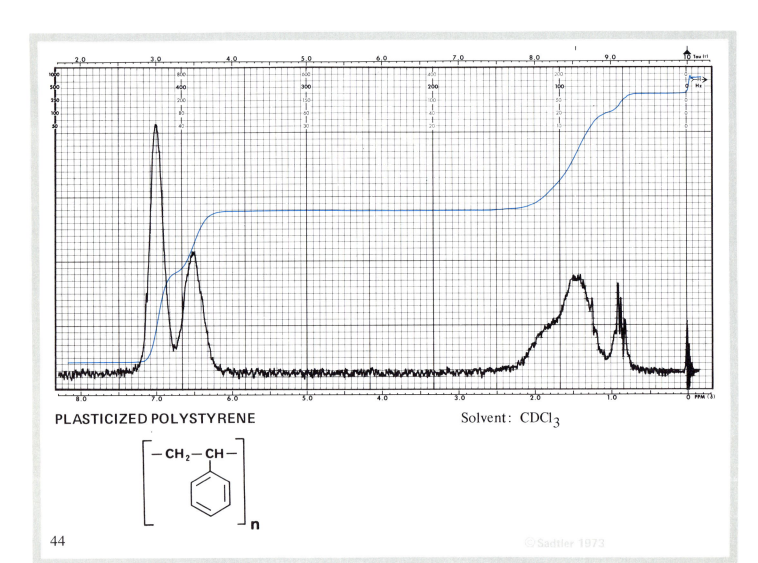

PLASTICIZED POLYSTYRENE

Solvent: $CDCl_3$

$$\left[-CH_2-CH- \right]_n$$

The intense bands at 2.05 and 5.38 ppm indicate that the major component of this sample is 1,4-poly-butadiene, with only a trace of the 1,2 isomer indicated by the weak band at 5.03 ppm. The styrene methylene and methine groups appear as a broad envelope of absorbance in the range from 1.1 to 2.5 ppm. The aromatic ring hydrogens appear as two bands at 6.55 and 7.05 ppm which suggests that the styrene units are grouped in blocks.

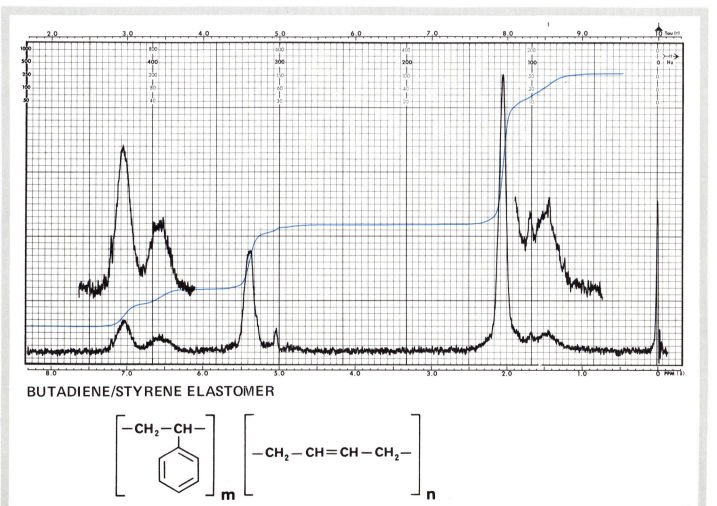

BUTADIENE/STYRENE ELASTOMER

45

The presence of polystyrene is indicated by the bands at 1.6 and 7.1 ppm. The fact that the styrene aromatic protons appear as one peak instead of two indicates that the styrene groups are separated by one or more butadiene units. The polybutadiene bands are similar to those of the spectrum on page 37, indicating that both the 1,4 and 1,2 butadiene units are present. Based upon the relative integration values, the composition of this copolymer is estimated to be, 21% polystyrene, 59% 1,2-polybutadiene and 20% 1,4-polybutadiene.

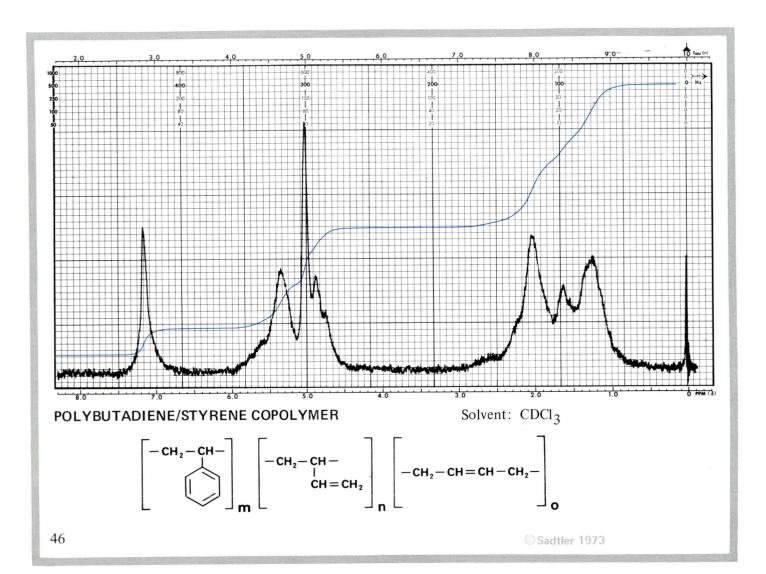

POLYBUTADIENE/STYRENE COPOLYMER   Solvent: $CDCl_3$

© Sadtler 1973

The characteristic absorption bands of cis-1,4-polyisoprene are observed at 1.70, 2.06 and 5.11 ppm. The weaker bands at 6.5 ppm represent the styrene ortho ring hydrogens and the band at 7.2 ppm represents the resonance of the meta and para hydrogens. The presence of another polyene unit in addition to 1,4-isoprene is suggested by the weak absorbance band at 4.67 ppm. The composition of this sample has been calculated to be 91% cis-1,4-polyisoprene and 9% polystyrene units.

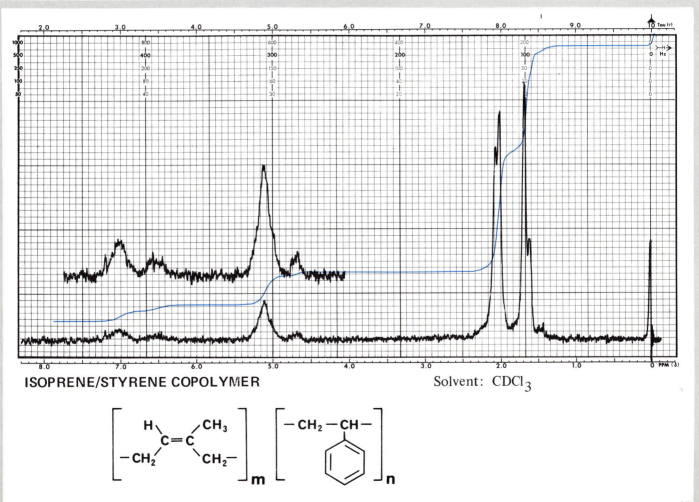

ISOPRENE/STYRENE COPOLYMER   Solvent: $CDCl_3$

47

THE VINYL POLYMERS

The vinyl polymers are structurally similar to the saturated hydrocarbon polymers which were discussed in the previous chapter, in that both groups of compounds are formed from unsaturated monomers which lose their double bonds during the polymerization process. The reaction is most frequently initiated by free radical catalysis.

Structurally, the vinyl polymers fall into two broad classes which possess either of the two following structures:

$$\left[-CH_2-\underset{\underset{X}{|}}{CH}- \right]_n \qquad\qquad \left[-CH_2-\overset{\overset{Y}{|}}{\underset{\underset{X}{|}}{C}}- \right]_n$$

VINYL   VINYLIDENE

where "X" may be chlorine, fluorine, a nitrile group or an oxygen containing functional group such as an ether, carboxyl alcohol, acetyl, etc. "Y" is usually either the same as "X" (the vinylidenes) or may be a methyl group (the methacrylates).

Included in this group of polymers are representatives in which the alternate X groups are joined together to form rings (the polyformals and butyrals), and certain hydrocarbon polymers which have been randomly chlorinated or chlorosulfonated.

Nomenclature

With only a few exceptions, these polymers follow the "vinyl/vinylidene" system of nomenclature. The major exceptions are polyacrylonitrile, the polyacrylates and the halogenated ethylenes such as polytetrafluoroethylene (PTFE).

A list of several monomers and their polymer names are presented in the table on the opposite page.

Monomer	Polymer	Page
Vinyl chloride	Polyvinyl chloride	54
Vinyl fluoride	Polyvinyl fluoride	— —
Vinyl acetate	Polyvinyl acetate	61
Methyl vinyl ether	Polyvinyl methyl ether	58
Methyl methacrylate	Polymethyl methacrylate	64
Vinylidene chloride	Polyvinylidene chloride	55
Acrylonitrile	Polyacrylonitrile	52
Tetrafluoroethylene	Polytetrafluoroethylene	— —

NMR Parameters

As a result of the presence of electron-withdrawing substituents attached to the carbon backbone, the residual polymer chain protons resonate over a much wider range than that observed for the hydrocarbon polymers.

The spectral bands produced by these polymers can be divided into three main categories; the $-CH_2-$ bands in the backbone, the $-CH-$ bands in the backbone and those which originate from proton containing groups on the attached side chains. In many cases, two or more of these bands may overlap creating difficulties in the interpretation of the spectrogram.

The following chart presents the chemical shifts of the $-CH_2-$ and $-CH-$ bands for some of the polymer types covered in this chapter.

$-CH_2-$	Polymer Type	$-CH-$	Page
1.6	Acetal	3.75 - 3.95	78
1.6	Ethers	3.45	58
1.7	Alcohol	4.05	56
1.82	Acetate	4.88	61
1.85	Carboxylic esters	—	63
2.1	Nitrile	3.15	52
2.14	Chloride	4.48	54

The solubility characteristics of the vinyl polymers vary depending upon the nature of the substituent. The vinyl ethers, acetates and methacrylates are normally readily soluble in the normal organic solvents at ambient temperature. Acrylonitrile, vinyl chloride and vinylidene chloride polymers usually require the use of dimethyl sulfoxide-d6 or ortho-dichlorobenzene at elevated temperatures, while the fluorinated polymers are usually insufficiently soluble for the preparation of a usable NMR spectrogram.

Commercial and common names

The listing presented below contains selected commercial and common names, and abbreviations which have been applied to the various polymers in this group. The numbers in parentheses refer to the list of manufacturers on page 275 to whom these trademark names belong.

Polyacrylonitrile
Acrilan (17), Orlon (1), Creslan (20), Zefran (21), polyvinyl cyanide, PAN.

Polyvinyl chloride
Breon (22), Darvic (9), Exon (14), Geon (23), Hostalit (24), Opalon (17), Pliovic (12), Poly-(monochloroethylene), PVC.

Polyvinylidene chloride (and high vinylidene content copolymers)
Polidene (25), Saran (3), Velon (14), Viclan (9), poly(1,1-dichloroethylene), PVDC, PVdC.

Polyvinyl alcohol
Cipoviol (26), Elvanol (1), Mowiol (24), Pevalon (27), Polyviol (28), PVA, PVAL, PVAl.

Polyvinyl ethers
Lutonal (29), Tetronic (32).

Polyvinyl acetate
Daratak (30), Edivil (31), Elvacet (1), Mowilith (24), Texicote (25), Vinavil (31), Vinnapas (28), PVA, PVAc, PVAC.

Polyacrylates and polymethacrylates
Acrylite (20), Aroset (33), Diakon (9), Lucite (1), Perspex (9), Plexiglas (34), PMA(polymethyl acrylate), PMMA (polymethyl methacrylate).

50

Polyvinyl pyrrolidone
Aldacol (35), Ganex (36), Kollidon (7), Polectron (36), Poly(N-vinyl-2-pyrrolidone), PVP.

Polyvinyl pyridines
Ionac PP series (37), Polysar látex 781 (38), Pliolite latex VP-100 (12).

Polyvinyl formal
Formvar (39), Pioloform F (28).

Polyvinyl butyral
Butvar (39), Pioloform B (28).

Chlorinated rubber
Alloprene (9), Parlon (5), rubber chloride.

Chlorinated polyethylene
Lutrigen (7), Tyrin (3).

Chlorosulfonated polyethylene
Hypalon (3).

VINYL POLYMERS

The methylene group appears as a broad band at 2.1 ppm and the methine absorbance is observed at about 3.15 ppm. The band at 2.52 ppm represents the solvent impurity DMSO-d5, the peak at 3.3 ppm probably represents a trace of water present in the solvent. The impurity peak at 3.72 ppm is of unknown origin.

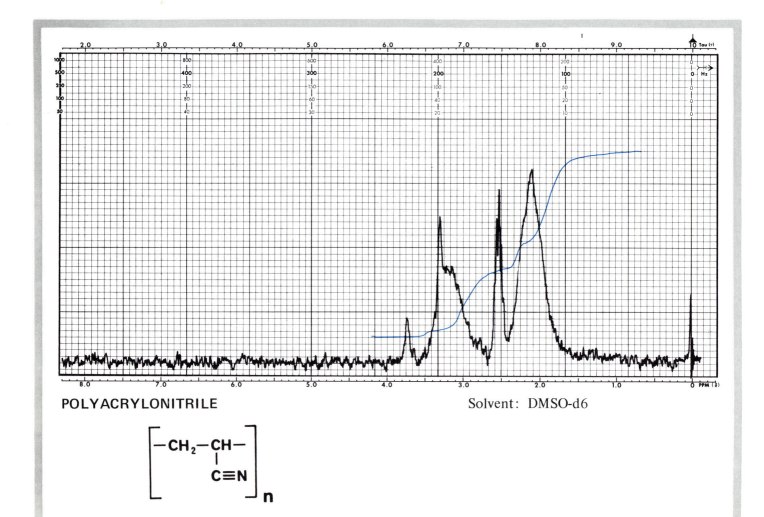

POLYACRYLONITRILE   Solvent: DMSO-d6

By law "Acrylic" fibers must contain no less than 85% acrylonitrile and the major absorption bands of this material at 2.1 and 3.13 ppm represent the methylene groups and methine groups respectively of polyacrylonitrile. The only evidence to suggest that a second polymer unit may be present is the single peak at about 3.32 ppm which may represent the methylene absorbance band of vinylidene chloride. The multiplet at 2.52 ppm represents the solvent impurity DMSO-d5.

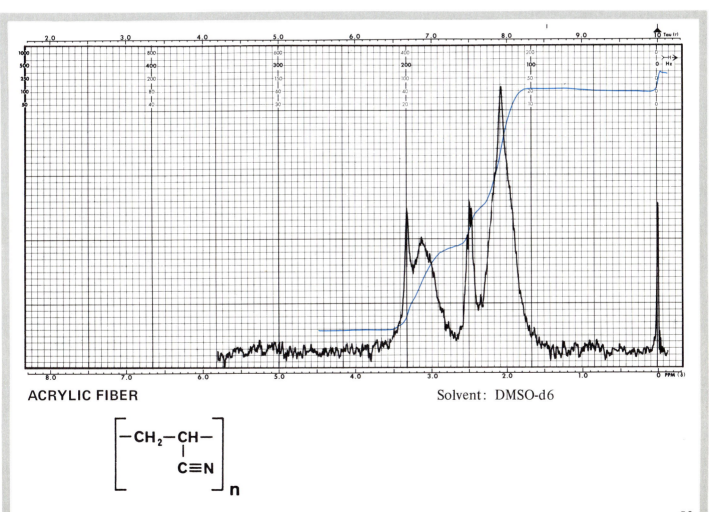

ACRYLIC FIBER

Solvent: DMSO-d6

$$\left[-CH_2-CH- \atop \quad\ C\equiv N \right]_n$$

53

VINYL POLYMERS

The methylene protons of polyvinyl chloride occur as a broad, complex absorbance band centered at about 2.14 ppm while the methine hydrogens which are strongly deshielded by the adjacent chlorine atoms resonate at about 4.48 ppm. Hot orthodichloro benzene was required to obtain a usable sample solution from this rather insoluble material.

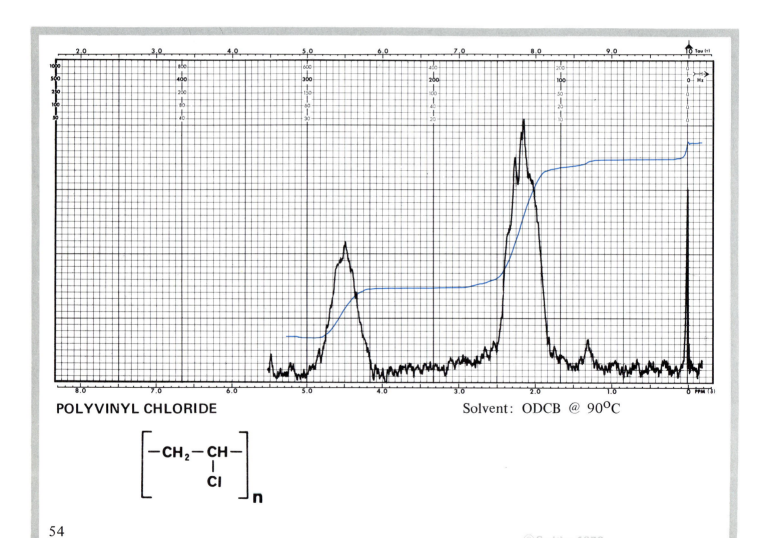

POLYVINYL CHLORIDE

Solvent: ODCB @ 90°C

$$\left[-CH_2 - CH - \atop | \atop Cl \right]_n$$

VINYL POLYMERS

In the NMR spectrum of this low molecular weight copolymer of vinyl chloride and vinylidene chloride, the band centered at about 2.3 ppm represents the methylene groups of the vinyl chloride units and the band at 4.6 ppm the methine protons of that homopolymer. The broad band near 3.0 ppm represents the methylene protons of the vinylidene chloride protons which are rather strongly deshielded due to their position beta to four chlorine atoms. The bands centered at about 3.72 ppm may represent a terminating group of the type Cl-CH$_2$-CH(Cl)-. The ratio of vinyl chloride units to vinylidene chloride units is about 2 to 1.

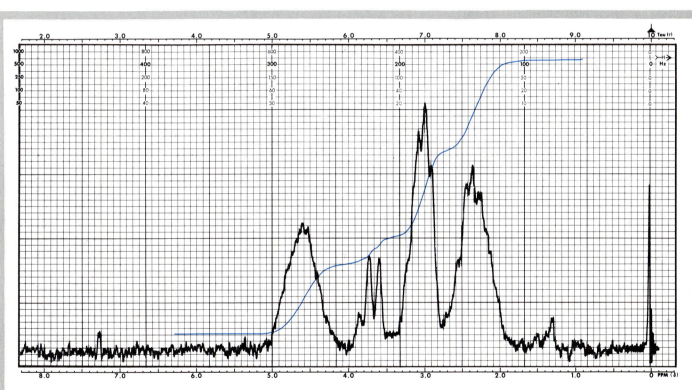

VINYL CHLORIDE/VINYLIDENE CHLORIDE COPOLYMER   Solvent: CDCl$_3$

$$\left[-CH_2-CH- \atop \qquad Cl\right]_m \left[-CH_2-C- \atop Cl\right]_n$$

VINYL POLYMERS

The NMR spectra of polyvinyl alcohols, which are prepared by the hydrolysis of polyvinyl acetates, often display absorbance bands representing residual acetate methyl groups. This band appears as a sharp singlet near 2.1 ppm. The polyvinyl alcohol methylene groups resonate at 1.7 ppm while the methine protons appear near 4.05 ppm. Based upon a comparison of the integration ratios, this sample of polyvinyl alcohol is found to contain about 10% residual acetate groups.

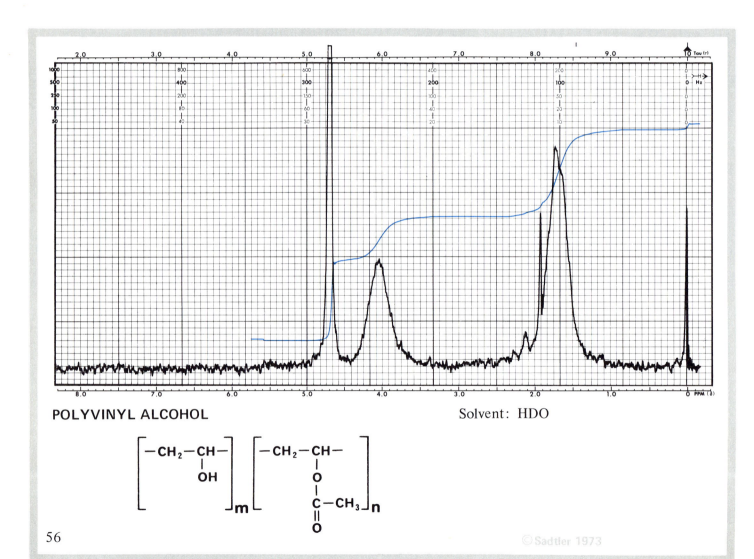

POLYVINYL ALCOHOL

Solvent: HDO

The polyvinyl alcohol methylene groups resonate as a broad, poorly resolved band centered at about 1.72 ppm. The methines are strongly deshielded by the adjacent hydroxyl groups and appear at much lower field, near 4.03 ppm. The sharp peak at 1.92 ppm probably represents a trace of acetone which is used in preparation of the polymer. The weak band at 2.11 ppm represents about 1 or 2% of residual polyvinyl acetate methyl groups.

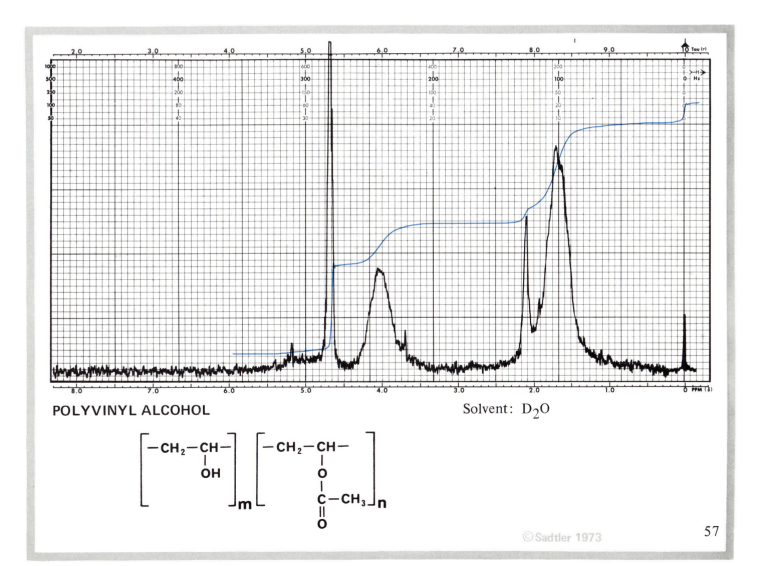

POLYVINYL ALCOHOL   Solvent: D_2O

57

VINYL POLYMERS

The strongest band in the NMR spectra of the polyvinyl methyl ethers arises from the methyl ether groups which resonate as a sharp singlet near 3.35 ppm. The methylene groups in the polymer backbone appear as a four line pattern centered at 1.7 ppm. The appearance of this four line pattern suggests that the two methylene protons are non-equivalent and represent the AB part of an ABX system. The methine protons occur as a broad multiplet near 3.45 ppm which overlaps with the methyl ether band.

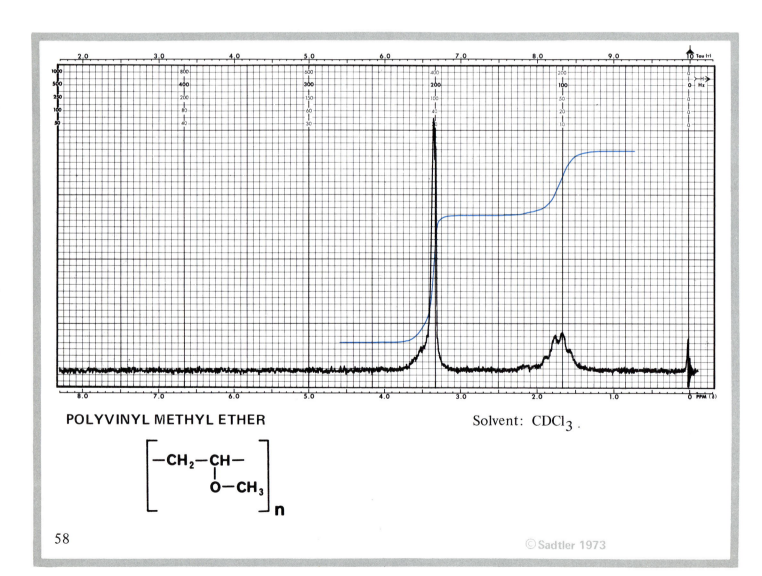

POLYVINYL METHYL ETHER

Solvent: CDCl$_3$.

$$\left[\begin{array}{c} -CH_2-CH- \\ \quad | \\ \quad O-CH_3 \end{array} \right]_n$$

58

The ethyl ether moiety appears as a triplet at 1.19 ppm and a broadened quartet centered at 3.48 ppm. The methylene groups of the polymer chain form a broad multiplet in the chemical shift range from 1.3 to 2.3 ppm. The methine protons resonate at lower field and overlap with the quartet of the ethyl group. The weak peaks at 2.3 and 7.09 ppm suggest that a trace of toluene is present in the sample solution as an impurity.

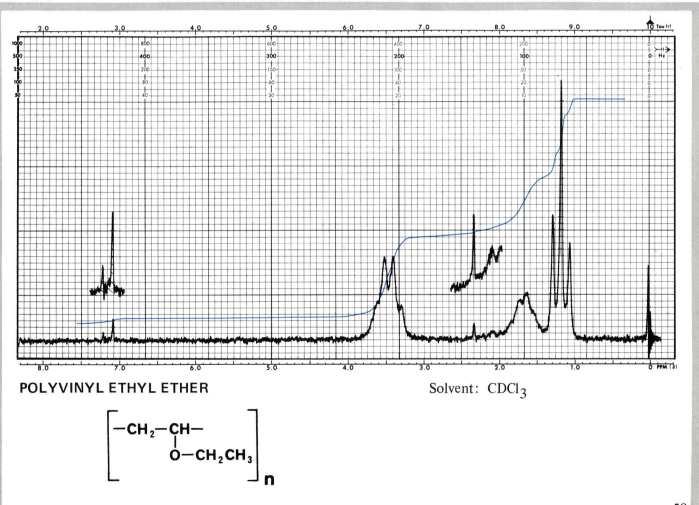

POLYVINYL ETHYL ETHER　　　　　　　　　　Solvent: $CDCl_3$

$$\left[-CH_2-\underset{\underset{O-CH_2CH_3}{|}}{CH}- \right]_n$$

59

VINYL POLYMERS

The isobutyl methyl groups appear as a six proton doublet at highest field near 0.91 ppm. The isobutyl methine and vinyl methylene groups overlap to form a complex band in the chemical shift range from 1.2 to 2.2 ppm. The isobutyl methylene groups which are bonded to the oxygen atoms resonate at lower field as a broadened doublet at 3.2 ppm while the vinyl methine protons which are also bonded to oxygen atoms appear at about 3.5 ppm as a broad, weak band. Polyvinyl isobutyl ether is used in the manufacture of adhesives, surface coatings, plasticizers and lubricating oils.

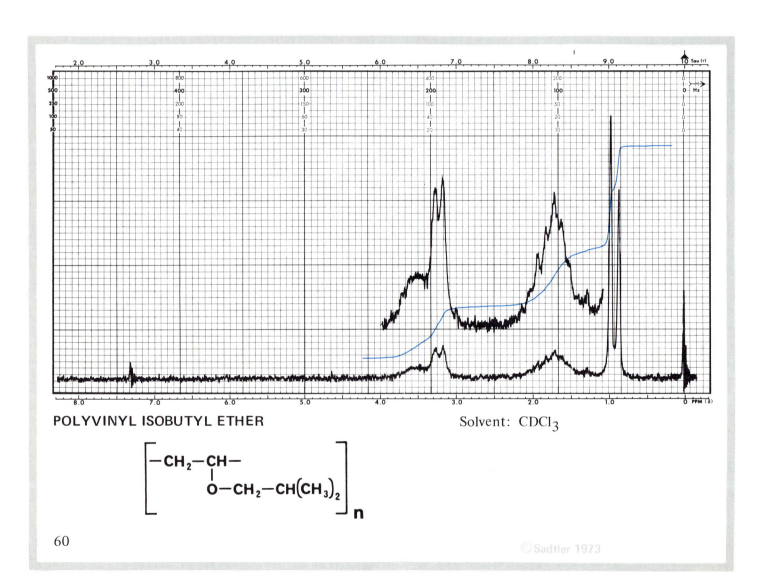

POLYVINYL ISOBUTYL ETHER Solvent: CDCl$_3$

$$\left[\begin{array}{l} -CH_2-CH- \\ \qquad\quad | \\ \qquad\quad O-CH_2-CH(CH_3)_2 \end{array} \right]_n$$

The acetate methyl groups resonate at 2.01 ppm as a slightly broadened singlet overlapping with the methylene groups which appear as a broad multiplet in the chemical shift range from 1.3 to 2.3 ppm. The strongly deshielded methines are observed as a broadened multiplet near 4.9 ppm. They resonate at lower field than the methine protons of polyvinyl methyl ether (page 58) due to the stronger deshielding effect of an ester functional group in relation to that of an ether group.

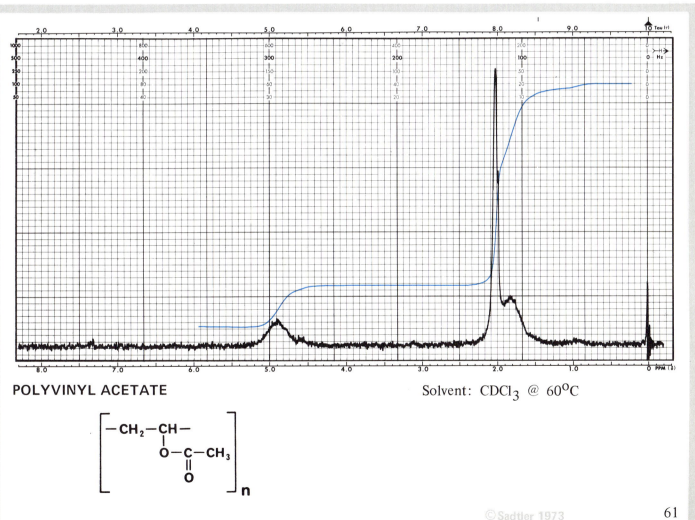

POLYVINYL ACETATE

Solvent: $CDCl_3$ @ $60^{\circ}C$

61

VINYL POLYMERS

The methyl ester groups are represented by the sharp single peak near 3.65 ppm. The ethyl ester groups display their characteristic high field triplet (1.26 ppm) and low field quartet patterns (4.12 ppm). The acrylic acid methylene and methine groups form a broad band of low intensity in the chemical shift range from 0.8 to 3.0 ppm. The composition of this sample is estimated to be: 55 mole % polyethyl acrylate, 45 mole % polymethyl acrylate.

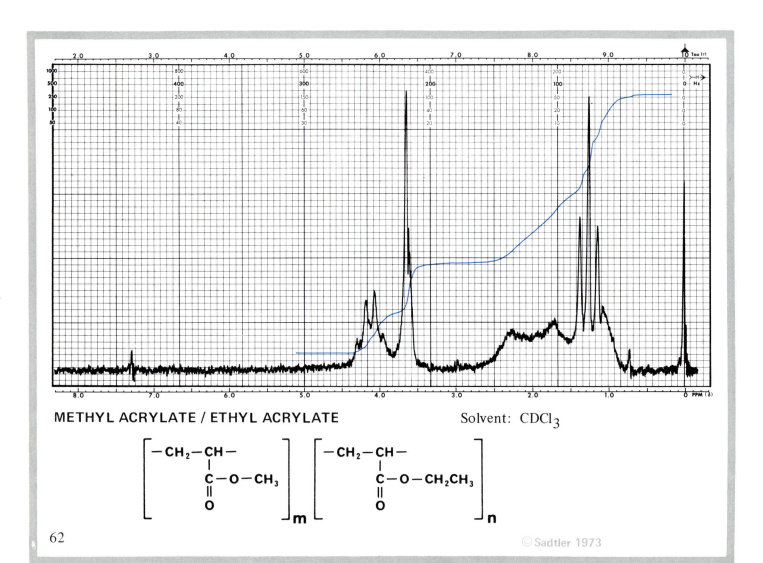

METHYL ACRYLATE / ETHYL ACRYLATE   Solvent: $CDCl_3$

The major features of this spectrum are due to the butyl ester groups. Their absorbance bands are located as follows: the methyl groups appear as a distorted triplet at 0.98 ppm, the two internal methylene groups overlap in the chemical shift range from 1.1 to 1.9 ppm, the methylene group adjacent to the ester moiety appears at 3.96. The acrylate methylene groups overlap with the butyl absorbance bands in the range from 1.1 to 1.9 while the acrylate methine group appears as a broad band with a chemical shift 2.18 ppm.

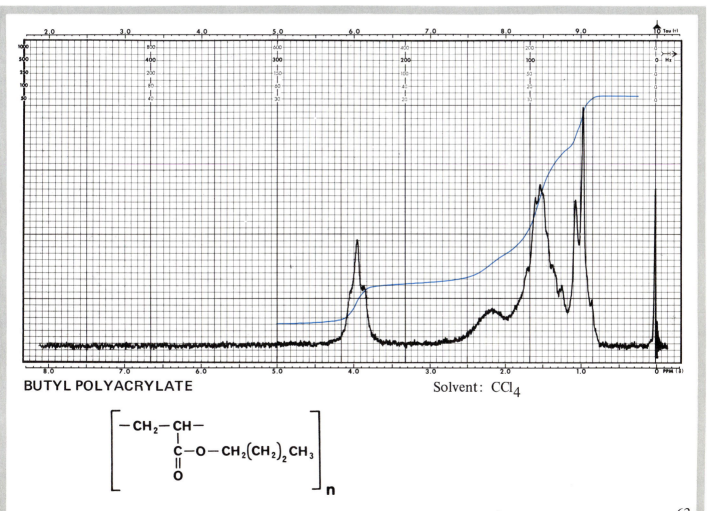

BUTYL POLYACRYLATE

Solvent: CCl$_4$

63

VINYL POLYMERS

The three peaks which appear at highest field represent methacrylate methyl groups of varying tacticity. The band at 0.90 ppm represents syndiotactic methyls, the band at 1.06 ppm arises from atactic methyl groups and the weak band at 1.25 ppm represents the resonance of the isotactic methyl groups. The corresponding isotactic, atactic and syndiotactic methylene groups absorb in the chemical shift range from 1.4 to 2.4 ppm. The methyl ester protons resonate at lowest field near 3.6 ppm.

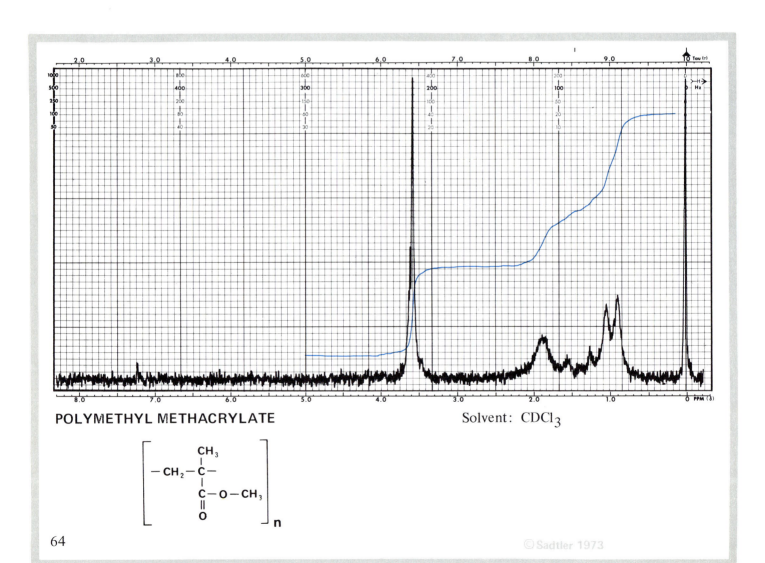

POLYMETHYL METHACRYLATE   Solvent: $CDCl_3$

The NMR spectrogram of this sample of polymethyl methacrylate is similar in general appearance to that on the previous page. The absence of a methyl peak at about 1.25 ppm indicates that this material contains few if any isotactic methyl groups. The ratio of syndiotactic to atactic methyl groups is similar to that of the previous sample, about 1 to 1.

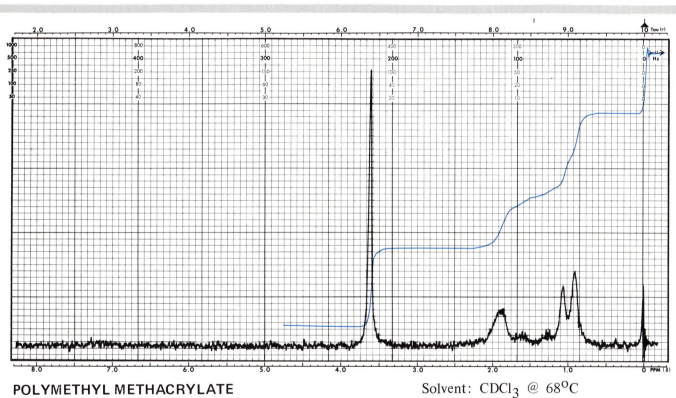

POLYMETHYL METHACRYLATE

Solvent: $CDCl_3$ @ 68°C

65

The ethyl ester groups are easily discerned by their characteristic triplet-quartet patterns with chemical shifts of 1.31 ppm and 4.07 ppm respectively. The methacrylate methylene groups form a broad, poorly resolved envelope of absorbance in the chemical shift range from 1.4 to 2.4 ppm. The absorbance band arising from methyl groups in syndiotactic triads appears at 0.92 ppm and the resonance due to methyl groups in atactic triads is observed as a broadened singlet at 1.06 ppm.

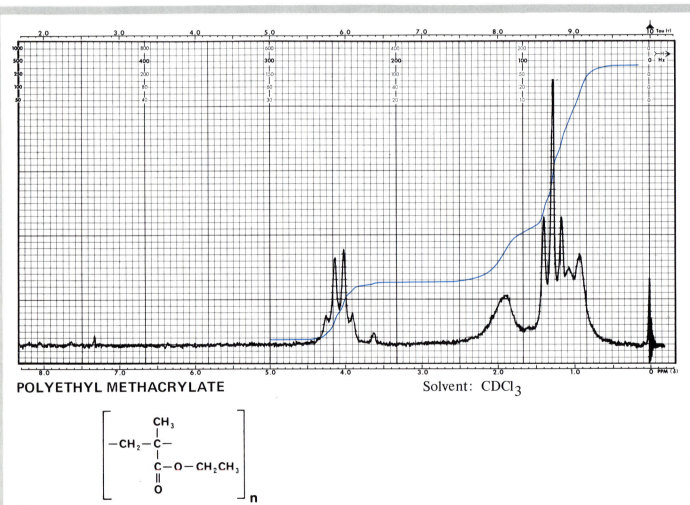

POLYETHYL METHACRYLATE

Solvent: $CDCl_3$

$$\left[-CH_2-\overset{\displaystyle CH_3}{\underset{\displaystyle \underset{\displaystyle O}{\overset{\displaystyle \|}{C}}-O-CH_2CH_3}{C}}- \right]_n$$

66

The high field area of this spectrum (0.5 ppm to 2.5 ppm) appears as a very complex pattern due to the overlap of five different hydrocarbon groups containing a total of twelve hydrogen nuclei. Only the CH_2 groups that are bonded to the oxygen atoms of the ester moieties are sufficiently deshielded to be clearly observed. These methylene groups appear as a broadened triplet at 3.96 ppm. At highest field, the terminal methyls of the butyl ester groups appear at about 0.95 ppm.

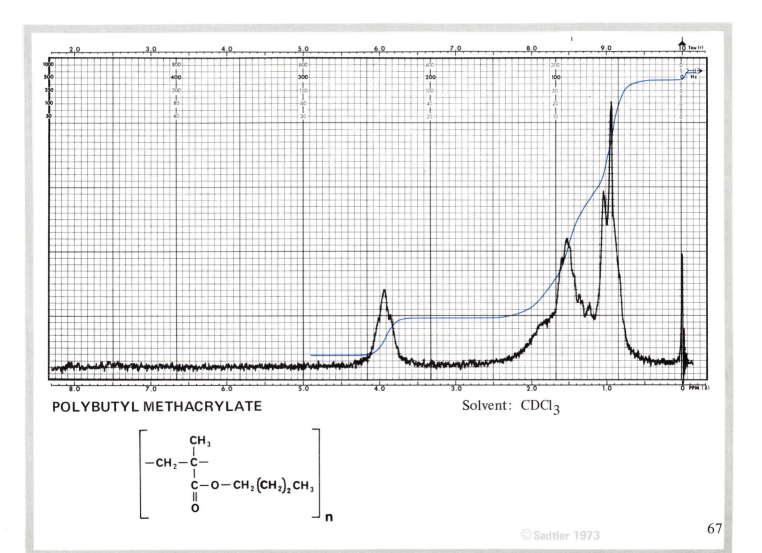

POLYBUTYL METHACRYLATE

Solvent: $CDCl_3$

67

The three methyl groups resonate at highest field producing a broadened doublet pattern centered at about 0.96 ppm. At slightly lower field, the chain CH_2 groups and isobutyl methine protons overlap to form a broad, unresolved band which is centered at about 1.94 ppm. At lowest field the isobutyl CH_2's which are bonded to the methacrylic acid oxygen atoms resonate as a very broadened doublet-like band at 3.76 ppm. The integration ratio of 2:3:9 is in good accord with the theoretical values for polyisobutyl methacrylate and is the only clear evidence to indicate the presence of the methacrylate methyl group.

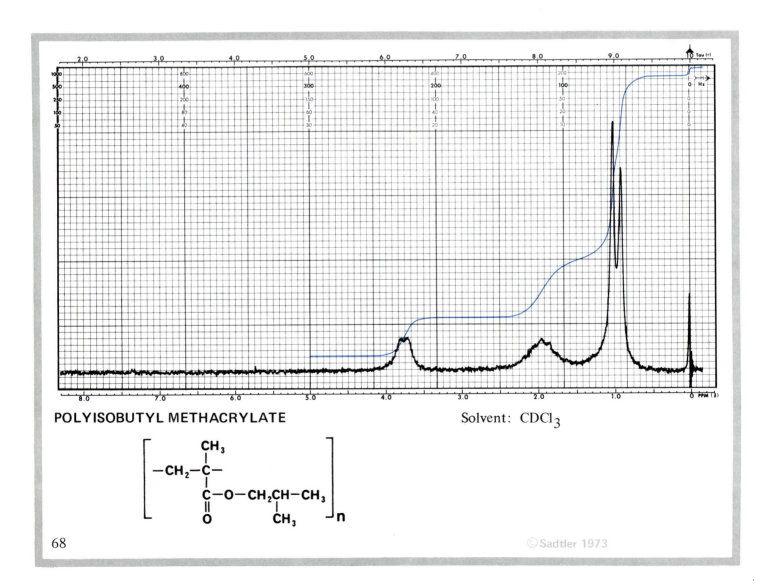

POLYISOBUTYL METHACRYLATE   Solvent: $CDCl_3$

Polyvinyl pyrrolidone, like most other polymers containing alicyclic rings, produces an NMR spectrogram characterized by very broad, poorly resolved absorbance bands. Based upon the relative integration values and chemical shifts, the three maxima may be assigned as follows: the band centered at 2.2 ppm represents the chain methylene group and the center methylene group of the pyrrolidone ring, the band near 3.3 ppm arises from the ring methylene group adjacent to the amide carbonyl, and the broad band centered at about 3.96 ppm represents the chain methine and ring methylene groups which are bonded to the nitrogen atom.

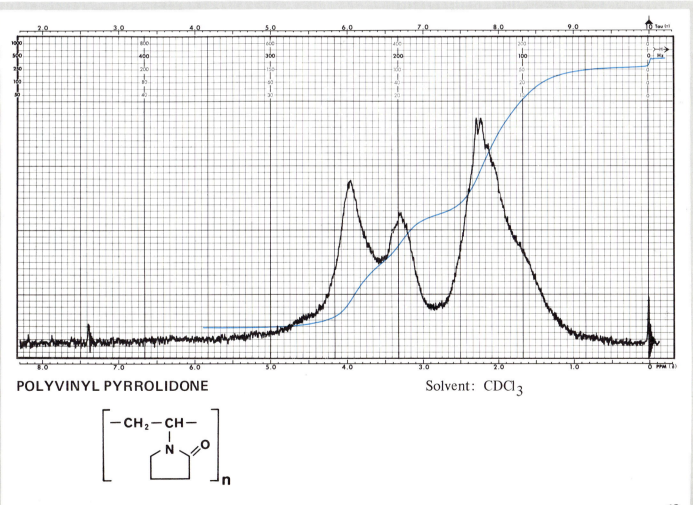

POLYVINYL PYRROLIDONE

Solvent: $CDCl_3$

69

The methylene and methine groups of the polymer chain overlap at high field to form a broad, single peak centered at about 1.82 ppm. The four pyridine ring hydrogens resonate at low field and display chemical shifts at slightly higher field than those of a reference compound such as 2-methyl pyridine, i.e.

	Polymer	2-methyl pyridine
H-3	6.79 ppm	7.03 ppm
H-4	7.12 ppm	7.43 ppm
H-5	6.35 ppm	6.96 ppm
H-6	7.69 ppm	8.42 ppm

The peak at 2.99 ppm arises from a trace of water present in the sample solution.

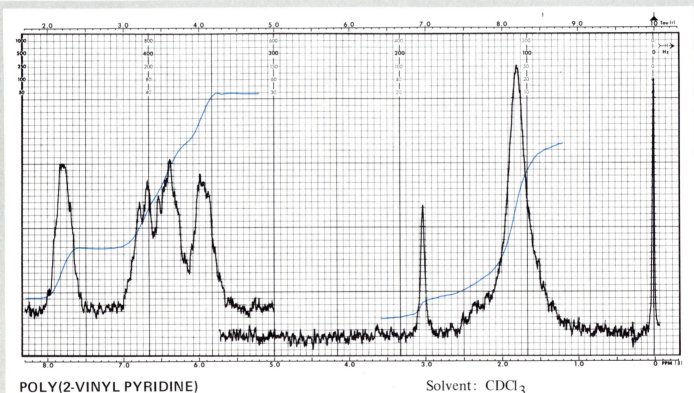

POLY(2-VINYL PYRIDINE)

Solvent: $CDCl_3$

$$\left[-CH_2-CH- \right]_n$$

The methylene and methine groups of both homopolymer units overlap to form a broad band in the chemical shift range from 1.2 to 2.3 ppm. The position-2 methyl group appears at 2.4 ppm as a broadened single peak. The seven different pyridine ring hydrogens form a complex series of poorly resolved bands in the chemical shift range from 6.0 to 8.5 ppm. The single peak at 3.21 ppm is exchangeable and probably represents water present in this hygroscopic polymer. The approximate composition of this sample has been calculated to be: 2-vinyl pyridine units 58%, 5-vinyl-2-methyl pyridine units 42%.

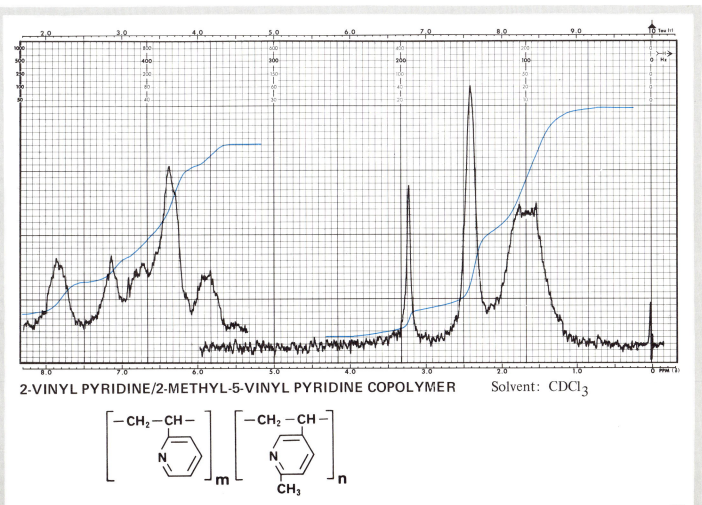

2-VINYL PYRIDINE/2-METHYL-5-VINYL PYRIDINE COPOLYMER   Solvent: $CDCl_3$

71

VINYL POLYMERS

In this spectrum of an ethylene/vinyl acetate copolymer, the acetate methyl groups resonate as a sharp singlet near 2.02 ppm while the methylene groups occur as a broad multiplet in the chemical shift range from 1.1 to 1.9 ppm. The methine protons resonate at lowest field near 4.88 ppm. The hydrogens of the polyethylene homopolymer appear as a sharp single peak at 1.28 ppm which overlaps with the absorbance bands of the PVA methylene protons. A comparison of the relative integration values indicates that the polymer is composed of 63% polyethylene units and 37% polyvinyl acetate units.

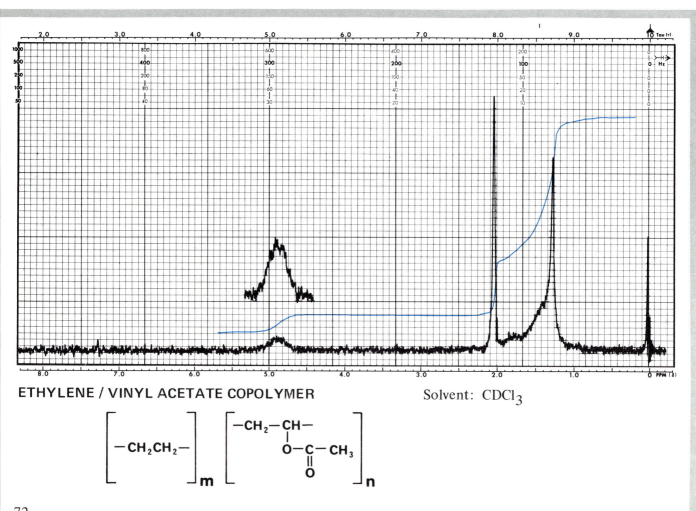

ETHYLENE / VINYL ACETATE COPOLYMER   Solvent: $CDCl_3$

$$\left[-CH_2CH_2-\right]_m \left[\begin{array}{c} -CH_2-CH- \\ | \\ O-C-CH_3 \\ \| \\ O \end{array}\right]_n$$

The increased intensity of the 1.29 ppm band of this ethylene/vinyl acetate copolymer indicates a higher percentage of polyethylene units than that observed in the previous spectrum. The chemical shifts of each component are practically identical to those observed in the spectra of the homopolymers (page 27 , page 61).

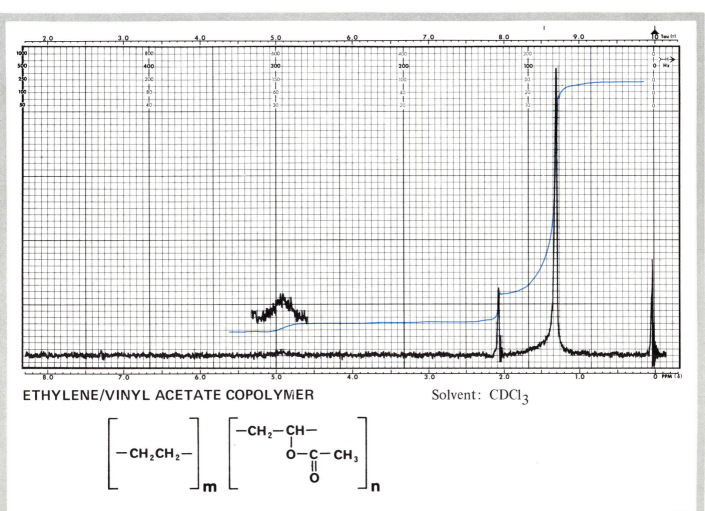

ETHYLENE/VINYL ACETATE COPOLYMER   Solvent: $CDCl_3$

73

The intensity of the 1.25 ppm band indicates a high percentage of ethylene units in this sample of an ethylene/ethyl acrylate copolymer. The only clear evidence to indicate the presence of ethyl acrylate units is the appearance of a clear quartet at 4.05 ppm. The remaining acrylate protons overlap at high field with the polyethylene protons (0.5 to 2.8 ppm). Based upon a comparison of the integration values, this copolymer is estimated to contain about 95% polyethylene units and 5% ethyl acrylate units.

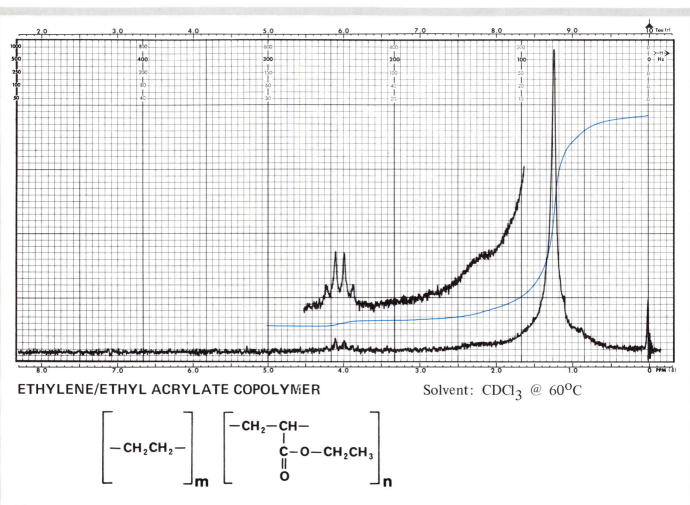

ETHYLENE/ETHYL ACRYLATE COPOLYMER   Solvent: $CDCl_3$ @ 60°C

74

In the NMR spectrogram of this copolymer of polyvinyl chloride and polyvinyl acetate, the methylene groups of each homopolymer overlap in the chemical shift range from 1.8 to 2.7 ppm. Similarly, the methine groups from the two different polymer units overlap to form a broad band in the chemical shift range from 4.0 to 4.9 ppm. The methyl groups of polyvinyl acetate appear as a sharp single peak at about 2.1 ppm. Based upon a comparison of the integration values, the composition of this copolymer has been calculated to be: 85% polyvinyl chloride units and 15% polyvinyl acetate units.

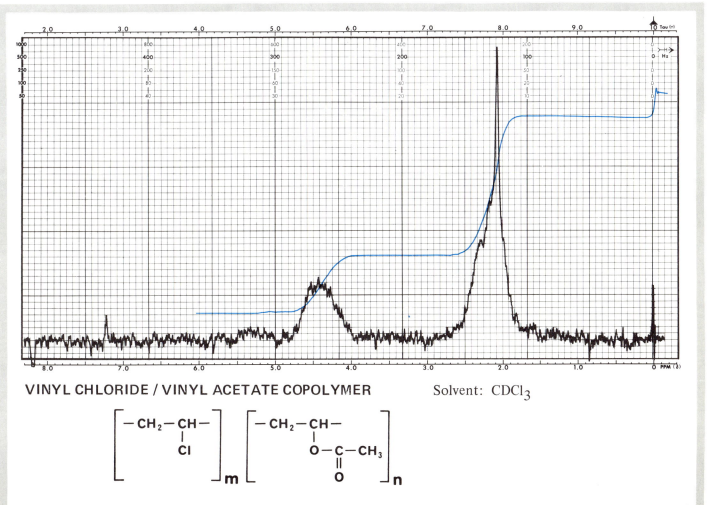

VINYL CHLORIDE / VINYL ACETATE COPOLYMER   Solvent: $CDCl_3$

75

The polystyrene aromatic protons appear as a single peak near 7.15 ppm indicating that the styrene units are isolated from each other by intervening methyl methacrylate units. The single peaks at 2.9 and 3.6 appear to represent two different types of methyl ester groups that are differentiated in chemical shift by their spatial position relative to the styrene aromatic rings. The methyl, methylene and methine groups which form the polymer chain form a complex, poorly resolved envelope of absorbance in the chemical shift range from 0.4 to 2.8 ppm. The ratio of methyl methacrylate units to styrene units is approximately 2 to 1.

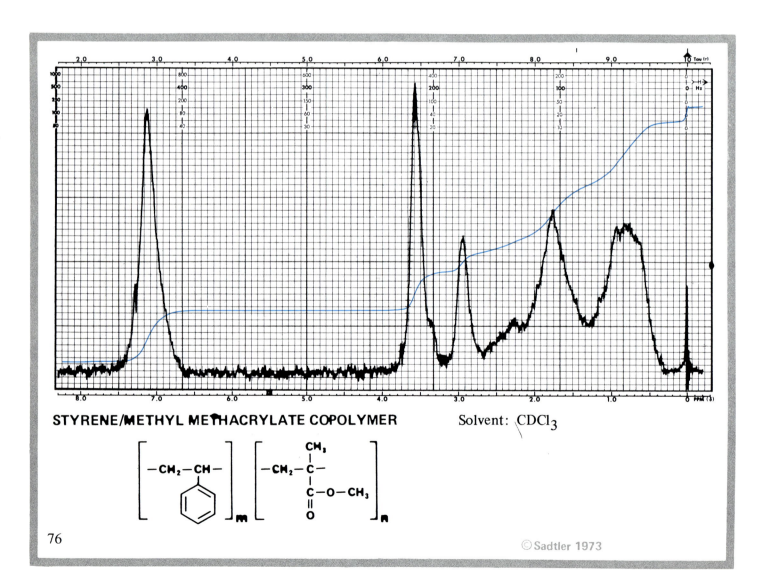

STYRENE/METHYL METHACRYLATE COPOLYMER   Solvent: CDCl₃

The characteristic absorbance bands at 2.02 ppm and 3.60 ppm indicate that the methyl methacrylate moiety is the major component of this copolymer. The weak triplet at 4.03 ppm represents the butyl ester of the CH_2-O- group. The sharp peak at 0.92 ppm probably represents the center peak of the butyl terminal methyl group. The methacrylate methyl peak at 1.12 ppm indicates that most of the methyl groups are present in atactic triads rather than syndiotactic or isotactic triads.

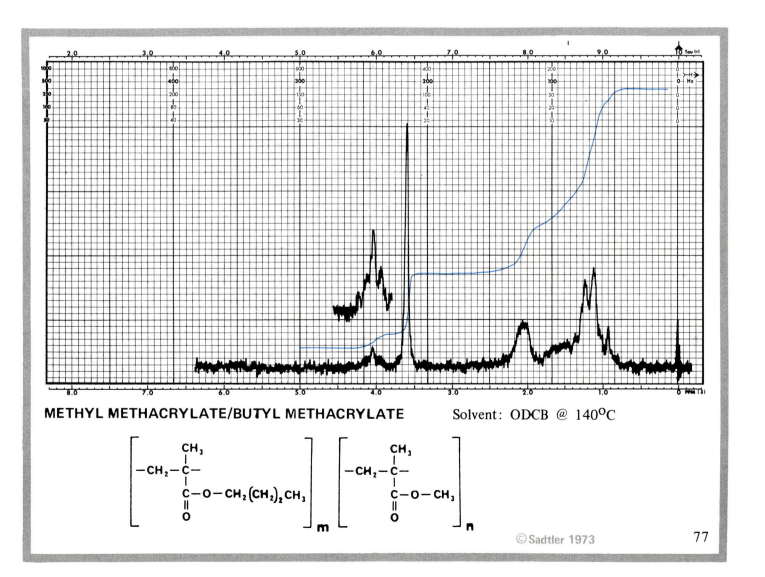

METHYL METHACRYLATE/BUTYL METHACRYLATE   Solvent: ODCB @ 140°C

77

VINYL POLYMERS

The reaction of formaldehyde with polyvinyl alcohol results in the formation of an alicyclic structure containing a methylene dioxy group (-O-CH$_2$-O) which resonates in the chemical shift range from 4.5 to 5.5 ppm as a complex series of peaks. The complexity of this absorbance band suggests that the two methylene protons are non-equivalent and appear to display geminal coupling. The presence of acetate methyl groups in the polyvinyl alcohol starting material is indicated by the sharp singlet near 2.0 ppm. The chain methylene groups resonate in the chemical shift range from 1.2 to 2.5 ppm while the methine protons appear at about 3.7 ppm.

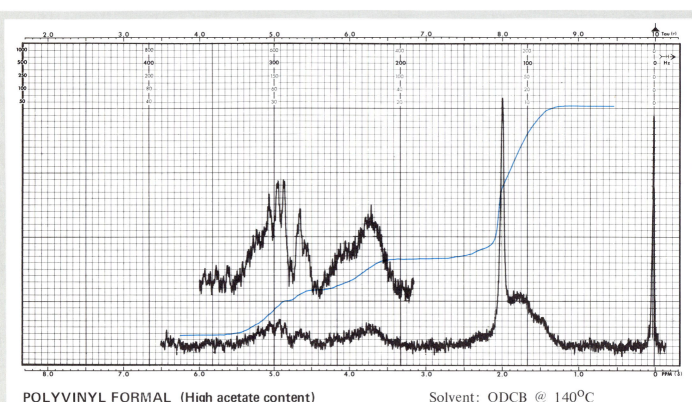

POLYVINYL FORMAL (High acetate content)   Solvent: ODCB @ 140°C

The dioxy-methylene groups ($-O-CH_2-O-$) which resonate at lowest field appear as a symmetrical four line pattern which appears to represent an AB pattern and indicates non-equivalence of these two methylene protons. The methylene groups in the chain resonate as a broad band centered at about 1.62 ppm while the chain methines produce a broad band in the range from 3.3 to 4.4 ppm. The sharp singlet at 2.03 ppm represents residual acetate methyl groups in the polymer.

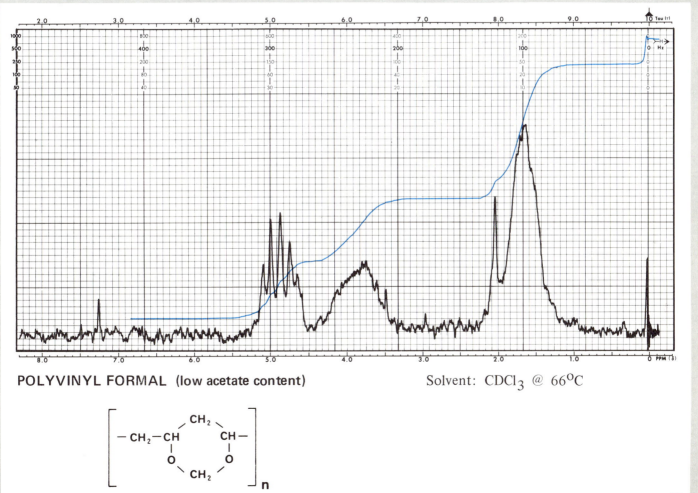

POLYVINYL FORMAL (low acetate content)   Solvent: $CDCl_3$ @ 66°C

VINYL POLYMERS

The cyclic methine in the environment (R-CH(OR)$_2$) appears as a weak band at lowest field, near 4.6 ppm. The broad band in the chemical shift range from 3.5 to 4.5 ppm represents all of the other hydrocarbon groups which are bonded to oxygen atoms. The vinyl and butyl methylene groups overlap in the chemical shift range from 1.1 to 2.3 ppm, while the butyl terminal methyl groups appear as a distorted triplet centered at 0.95 ppm.

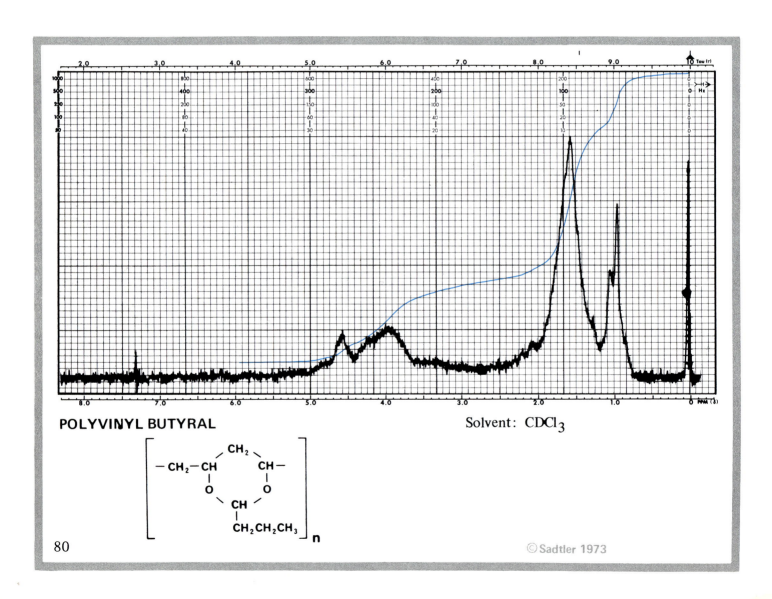

POLYVINYL BUTYRAL

Solvent: CDCl$_3$

The chlorination of natural rubber (polyisoprene) results in the formation of a saturated cyclic structure such as that represented below. The presence of a strongly deshielding substituent such as chlorine in the polymer results in chemical shifts in the 3.5 to 5.5 ppm region for the methine groups adjacent to the the chlorine atoms. The proton groups which are not substituted by chlorine resonate at higher field (0.8 to 3.5 ppm). The spectra of unchlorinated polyisoprene and natural rubber appear on pages 39 and 40 respectively.

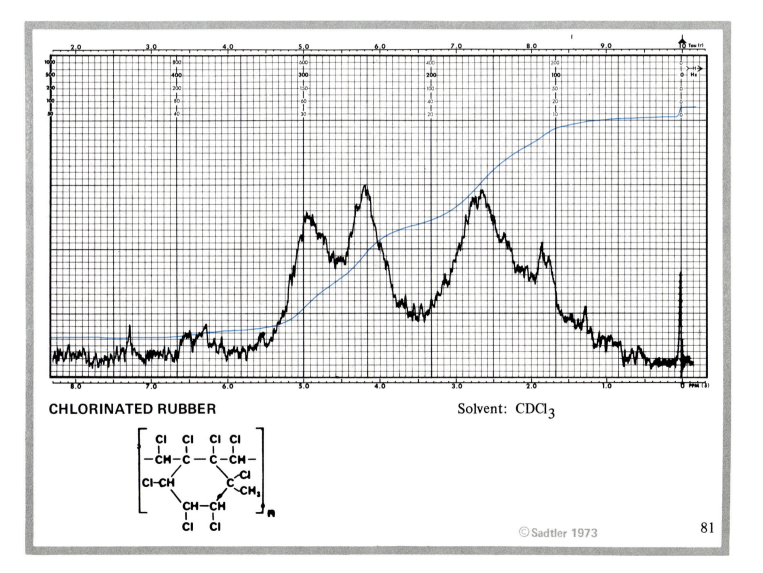

CHLORINATED RUBBER

Solvent: CDCl$_3$

©Sadtler 1973

81

VINYL POLYMERS

This sample of chlorinated polyethylene produces a spectrogram similar in general appearance to that of chlorinated rubber on page 81. The complete absence of the characteristic polyethylene $(CH_2)_n$ band at about 1.3 ppm indicates a high degree of chlorine substitution in the polymer chain. The broad band at about 2.8 ppm represents chain methylene groups which are beta to one or more chlorine atoms while the band centered at about 4.5 ppm represent methine protons which are alpha to one and beta to one or more chlorine atoms.

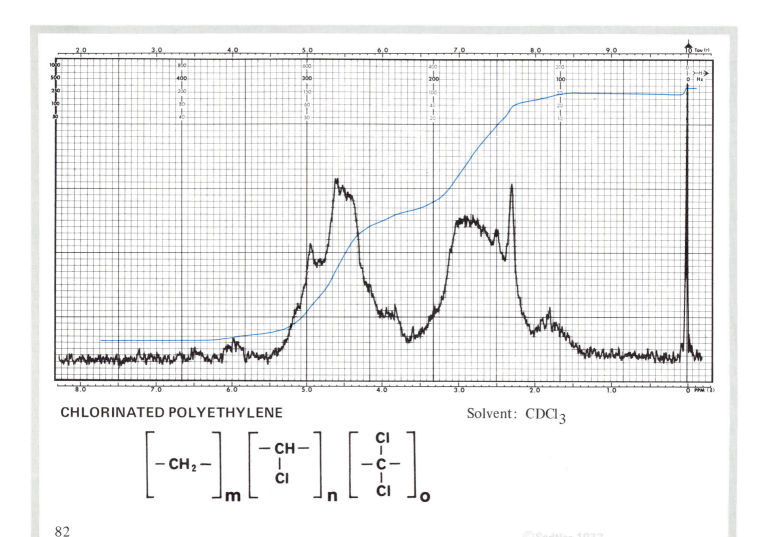

CHLORINATED POLYETHYLENE

Solvent: $CDCl_3$

$$\left[-CH_2-\right]_m \left[\begin{array}{c} -CH- \\ | \\ Cl \end{array}\right]_n \left[\begin{array}{c} Cl \\ | \\ -C- \\ | \\ Cl \end{array}\right]_o$$

In most chlorosulfonated polyethylenes, the ratio of chlorine substitution to chlorosulfonate substitution is about 30 to 1, and thus there is little evidence observed in their spectra to indicate the presence of chlorosulfonate groups. The weak band at 3.9 ppm represents the methines bonded to chlorine. The sharp peak at 1.26 represents the methylene groups of relatively long segments of unsubstituted polymer. The broad band centered at about 1.6 ppm represents those methylene groups beta or gamma to a chlorine substituent (CH$_2$-CH(Cl)- and CH$_2$-C-CH(Cl)).

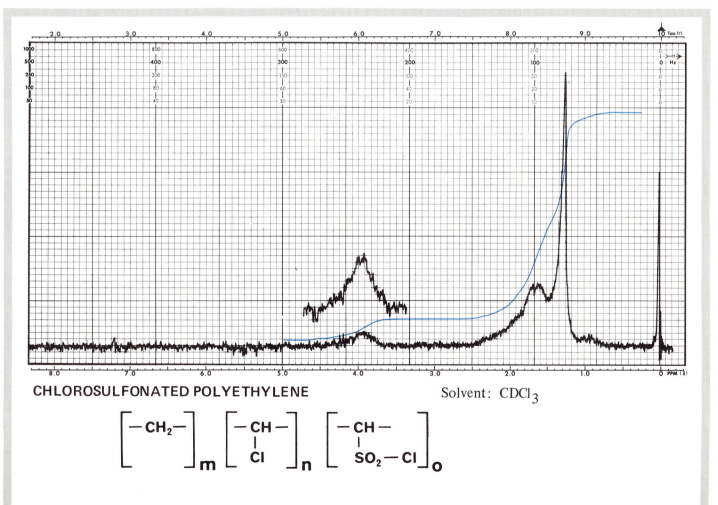

CHLOROSULFONATED POLYETHYLENE Solvent: CDCl$_3$

$$\left[-CH_2-\right]_m \left[\begin{array}{c}-CH-\\ |\\ Cl\end{array}\right]_n \left[\begin{array}{c}-CH-\\ |\\ SO_2-Cl\end{array}\right]_o$$

83

In comparison to the previous sample, this spectrogram shows a diminuation of the polyethylene band at 1.3 ppm and intensification of the 1.8 and 4.0 ppm bands indicating a much higher degree of chlorine substitution in this polymer. Although formed by completely different reactions, the molecular structure and NMR spectrum of chlorosulfonated polyethylene is similar in many respects to that of polyvinyl chloride on page 54 .

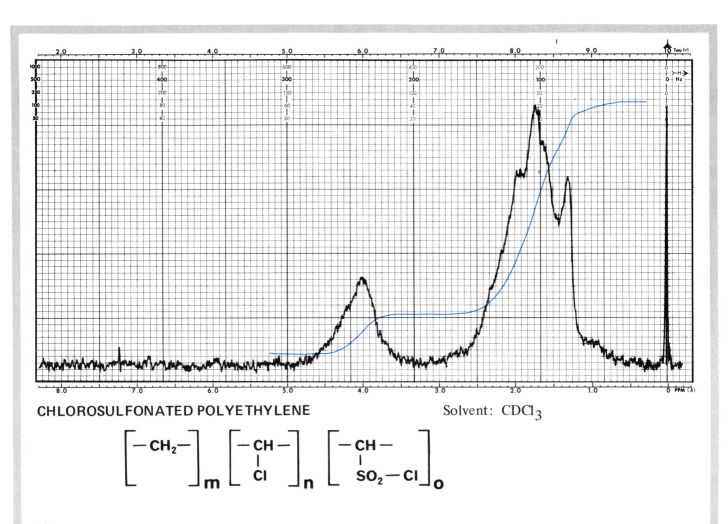

CHLOROSULFONATED POLYETHYLENE   Solvent: $CDCl_3$

$$\left[-CH_2-\right]_m \left[\begin{array}{c}-CH-\\ \ \ |\\ \ \ Cl\end{array}\right]_n \left[\begin{array}{c}-CH-\\ \ \ |\\ SO_2-Cl\end{array}\right]_o$$

THE POLYETHERS

The polyethers (or polyglycols) are the first group of "condensation" polymers to be discussed thus far and are distinguished from the previous _addition_ polymers by the fact that their backbone chain _does not_ consist of an uninterrupted chain of carbon atoms. In this series, a relatively simple hydrocarbon fragment alternates with an ether oxygen linkage. The hydrocarbon fragment may contain one (polyoxymethylene), two (polyethylene glycol), three (polypropylene glycol) or more carbon atoms, or may be an aromatic ring such as that of the polyphenylene oxides.

The simple aliphatic polyethers are obtained by the acid or base catalyzed polymerization of the appropriate cyclic ether or diol. The aromatic polyethers are synthesized by an unusual oxidative coupling reaction of their phenolic starting materials. Normally, the name of the polymer follows that of the monomer (or apparent monomer) although certain distinctions are sometimes made based upon the molecular weight or physical state of the final product. For example, low molecular weight polymers containing the formal $(-O-CH_2-)$ repeating unit are usually referred to as polyoxymethylenes but are called polyformaldehydes or acetal resins when obtained as tough, high molecular weight plastics. Similarly, polymers containing ethylene glycol $(O-CH_2CH_2-)$ units are called polyethylene glycol or polyoxyethylene when they are low molecular weight liquids or waxy solids but are referred to as polyethylene oxides when they are plastics capable of being extruded in sheet form.

Two other molecular variations are considered in this group; the polysulfones which contain a sulfone $(-SO_2-)$ linkage in addition to an ether bond, and the polysulfides in which one or more sulfur atoms replace some of the oxygen linkages along the chain.

The polysulfones are usually prepared by the condensation of the disodium salt of bisphenol-A with 4,4'-dichlorodiphenyl sulfone. The polysulfides are prepared by the reaction of dihalides such as bis-chloroethyl formal $(Cl-CH_2CH_2-O-CH_2-O-CH_2CH_2-Cl)$ with sodium polysulfide.

All of the polymers of this polyether group are formed by condensation reactions which are made possible by the difunctionality of their monomers.

NMR Parameters

With the exception of the polyphenylene oxides and the polysulfones, all of the polymers of this group possess methylene groups which are bonded to oxygen $(-O-CH_2-R)$. These methylene ether groups resonate over a fairly narrow chemical shift range from 3.4 to 3.9 ppm for the polypropylene glycols and polysulfides respectively. The dioxymethylene (formal) groups $(-O-CH_2-O)$ appear in the NMR spectra as a sharp singlet in the 4.6 - 4.7 ppm range; the methylene groups bonded to sulfur occur in the 2.8 - 2.9 ppm range; the isopropylidene methyl groups of bisphenol-A resonate as a sharp singlet near 1.7 ppm while the aromatic protons from this intermediate resonate over the chemical shift range from 6.5 to 7.8 ppm.

Other bands which may appear in the spectra of the polyethers will be characteristic of the specific monomer which has been employed, such as the doublet at 1.13 ppm which is observed for the branched methyl group of propylene glycol.

The low molecular weight varieties of the polyethers are soluble in both water and organic solvents. The degree of water solubility decreases with increasing hydrocarbon fragment size and with increasing molecular weight of the polymer.

The polyethers are frequently used as intermediates in the preparation of surface active agents and as pre-polymers in the synthesis of polyurethanes.

Commercial Names

Polyoxymethylene
    Celcon (40), Delrin (1), polyacetal, acetal resin, polyformaldehyde.

Polyethylene glycol
    Carbowax (41), Gafanol (36), Poly G (42), Polyox (41), polyoxyethylene, polyethylene oxide, polyglycol, PEG.

Polypropylene glycol
    Desmophen (43), Niax (41), Thanol F-3002 (44), polyoxypropylene, polypropylene oxide, PG, PPO, POP.

Polytetramethylene glycol
    Polymeg (45), poly(1,4-butylene glycol).

Polyphenylene oxide
    Arylon (10), Noryl (46), poly(2,6-dimethyl-1,4-phenylene oxide), polyxylenol, PPO.

Polysulfone -- Thermalux (47).

Polysulfide -- Dion (48), Thiokol (49).

Although polyformaldehyde is insoluble in all normal NMR solvents, the polymer can be decomposed by trifluoroacetic acid producing reaction products which appear in the spectrum below. The proportions of these bands gradually change with time as the reaction continues, the higher field peaks becoming smaller and the lower field bands become more intense. Breaking of polyether and polysulfide chains with gradual esterification of the terminal hetero atoms is characteristic of this strong acid.

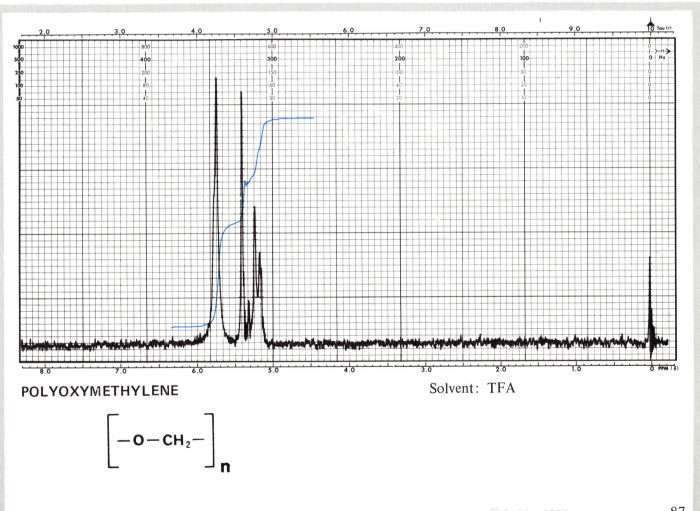

POLYOXYMETHYLENE

Solvent: TFA

$$\left[-O-CH_2- \right]_n$$

87

POLYETHERS

Polyethylene glycol, which is prepared by the condensation of ethylene oxide or ethylene glycol, produces an NMR spectrogram consisting of one single, sharp peak which represents all of the methylene groups in the chain. In the spectra of the high molecular weight materials, the terminal hydroxyl protons are present at such a small concentration that their resonance is usually not observed. Polyethylene glycol is widely used in the preparation of plasticizers and surface active agents such as the sample scanned on page 89.

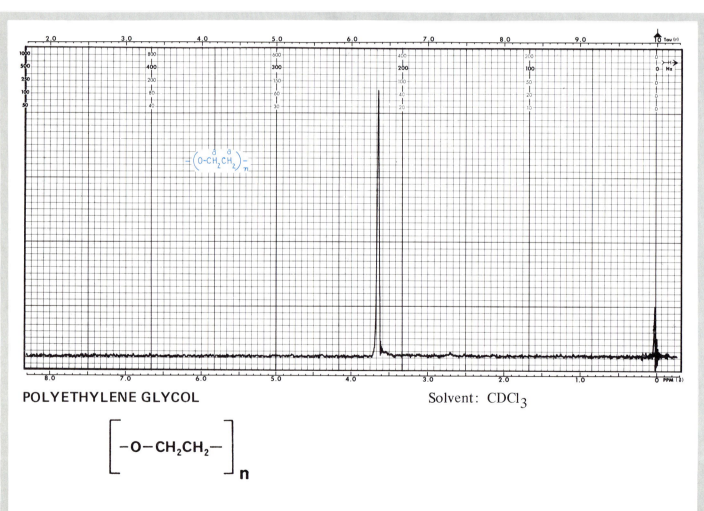

POLYETHYLENE GLYCOL

Solvent: $CDCl_3$

$$\left[-O-CH_2CH_2- \right]_n$$

In addition to the polyethylene glycol band at 3.65 ppm, the NMR spectrogram of this surface active agent contains bands due to the stearic acid moiety. These bands are observed at 0.89 ppm (the terminal methyl group), at 1.28 ppm (the chain methylene groups), and 2.34 ppm (the methylene bonded to the carbonyl). The band at 2.72 ppm probably represents terminating hydroxyl groups, while the triplet at 4.25 ppm represents the resonance of the ethylene glycol methylene group bonded to the ester oxygen atom. A comparison of the integration values indicate that the average molecule contains about 10 ethylene oxide units.

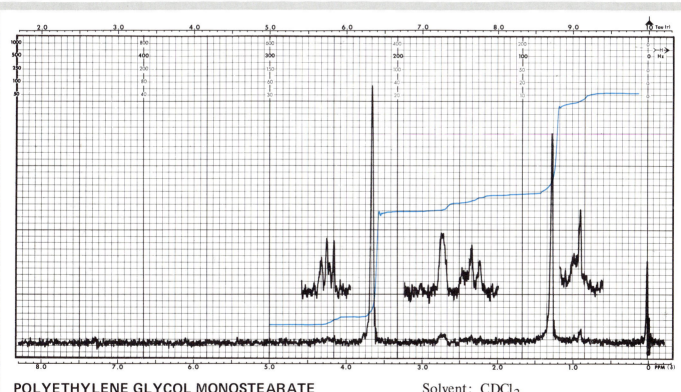

POLYETHYLENE GLYCOL MONOSTEARATE   Solvent: $CDCl_3$

$$CH_3(CH_2)_{15}CH_2 - \overset{\overset{\displaystyle O}{\|}}{C} - O - CH_2 - CH_2 (O - CH_2 - CH_2)_m - OH$$

89

The branching methyl groups resonate at 1.14 ppm as a doublet due to coupling to the adjacent methine protons. The methylene and methine groups which are bonded to the strongly deshielding oxygen atoms overlap to form a complex multiplet in the chemical shift range from 3.2 to 4.0 ppm. The molecular weight of this sample is about 2000 molecular weight units and so the average molecule would contain about 34 propylene oxide units.

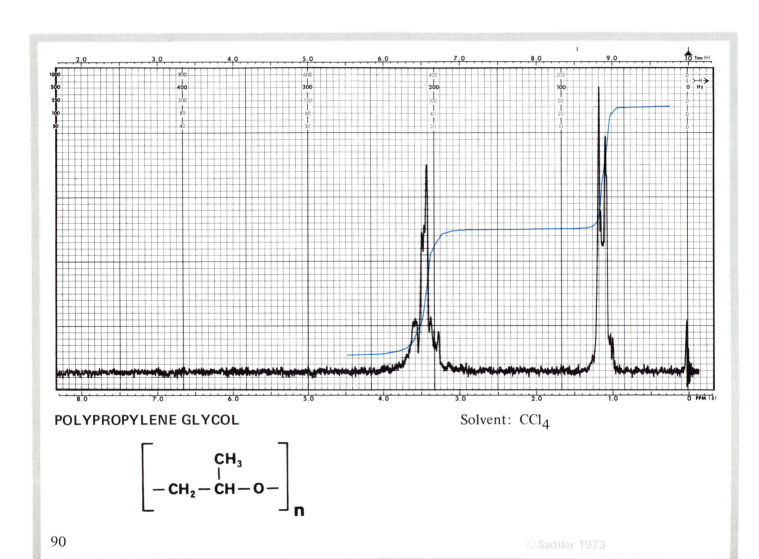

POLYPROPYLENE GLYCOL

Solvent: CCl_4

$$\left[-CH_2-\overset{\overset{\displaystyle CH_3}{|}}{CH}-O- \right]_n$$

This sample possesses an average molecular weight of about 400 and thus each molecule contains about seven propylene glycol units. The fine structure observed in this spectrum, in comparison to that on the previous page, is due to the fact that the terminal propylene oxide groups possess slightly different chemical shifts from the internal groups and are more noticeable in a low molecular weight sample such as this.

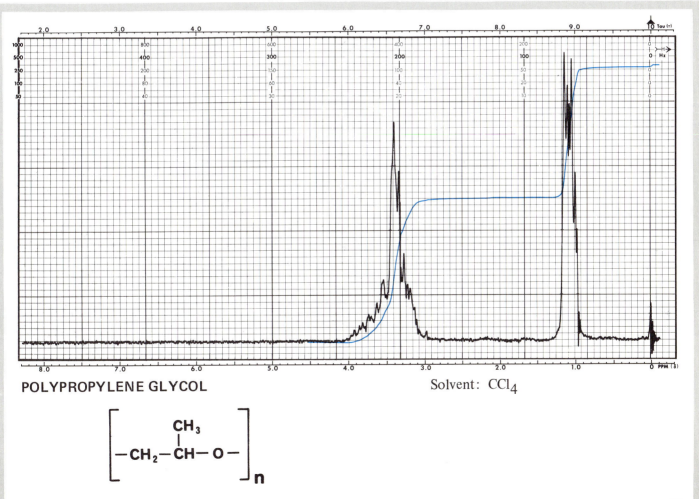

POLYPROPYLENE GLYCOL

Solvent: CCl_4

$$\left[-CH_2-\underset{\underset{CH_3}{|}}{CH}-O- \right]_n$$

91

The two central methylene groups resonate at high field (1.63 ppm) as a distorted triplet and the two methylene groups bonded to oxygen atoms appear at lower field (3.4 ppm). The characteristic distortion of these two triplets is probably due to virtual coupling and it is observed in the NMR spectra of all 1,4-disubstituted butane groups. The weak singlet at 2.61 ppm is due to exchangeable protons and probably represents terminal hydroxyl groups in addition to any water which may be present in this sample.

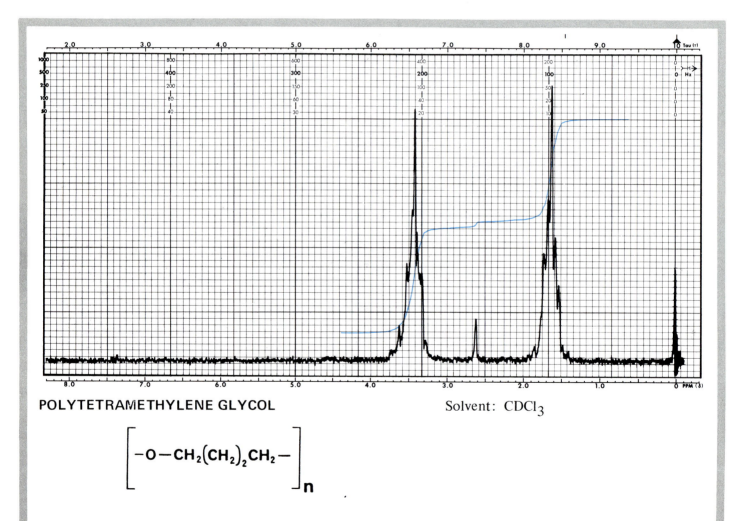

POLYTETRAMETHYLENE GLYCOL

Solvent: $CDCl_3$

$$\left[-O-CH_2(CH_2)_2CH_2- \right]_n$$

The methylene and methine groups which are bonded to the ether oxygen atoms possess similar chemical shifts and overlap to form a complex multiplet centered at about 3.35 ppm. The ethyl branches are represented by a poorly resolved pentet near 1.5 ppm and a clear triplet at 0.92 ppm. A very weak exchangeable proton band occurs at about 2.74 ppm and probably represents terminal hydroxyl groups along with any water present in the sample solution.

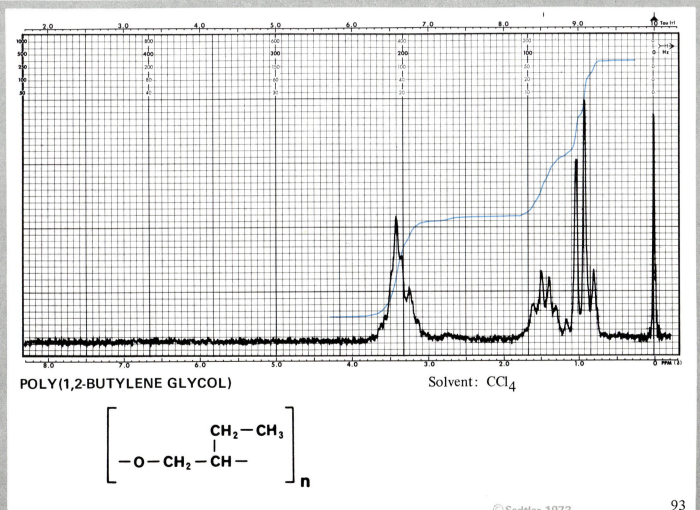

POLY(1,2-BUTYLENE GLYCOL)   Solvent: CCl_4

$$\left[-O-CH_2-\underset{\underset{CH_2-CH_3}{|}}{CH}- \right]_n$$

93

This copolymer of ethylene glycol and propylene glycol displays the characteristic absorbance bands of both polymers. The ethylene glycol units result in the intense sharp singlet at 3.64 ppm while the propylene glycol units are represented by the doublet at 1.15 ppm and the weak broadened band near 3.48 ppm. Based upon the relative integration values, the ratio of ethylene glycol units ot propylene glycol units is estimated to be 5.4 to 1.0 (about 15 mole % PPG, 85 mole % PEG).

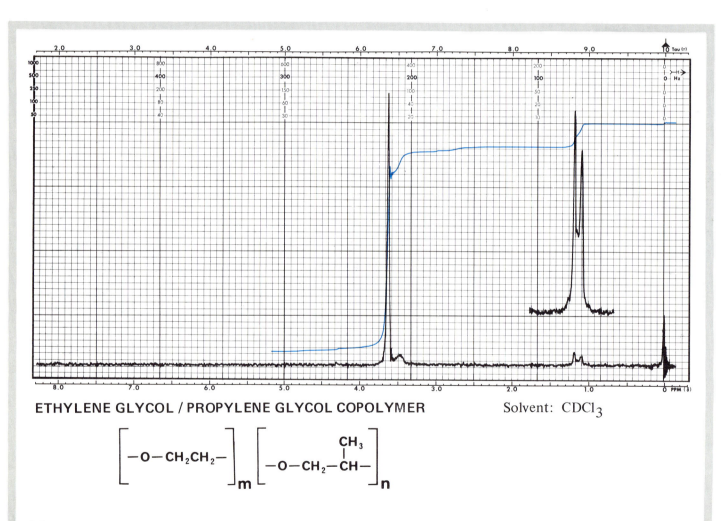

ETHYLENE GLYCOL / PROPYLENE GLYCOL COPOLYMER   Solvent: CDCl₃

$$\left[-O-CH_2CH_2- \right]_m \left[\begin{array}{c} CH_3 \\ | \\ -O-CH_2-CH- \end{array} \right]_n$$

The two equivalent methyl groups resonate at highest field (2.11 ppm) as a single peak. The aromatic protons appear as a single, two hydrogen peak at 6.5 ppm which is slightly broadened by long range coupling to the methyl protons. Although other polymers contain methyl groups which resonate at or near 2.1 ppm, few of them display a singlet at 6.5 ppm such as this.

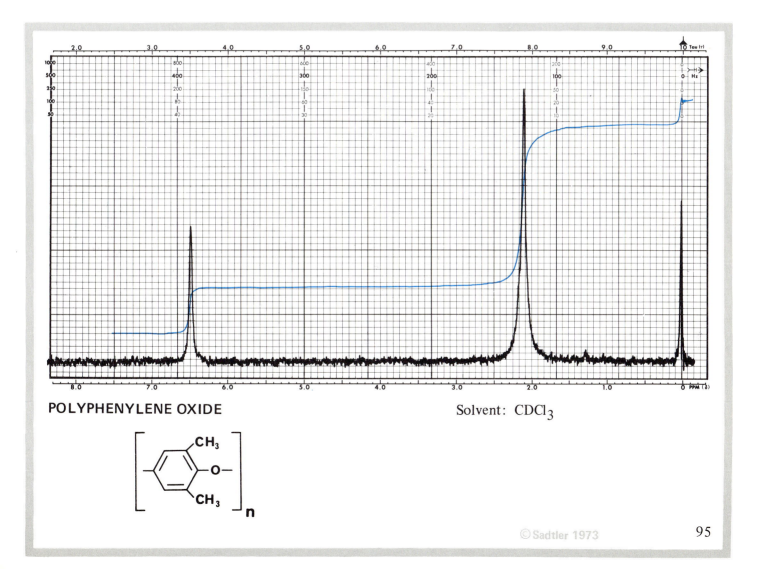

POLYPHENYLENE OXIDE

Solvent: CDCl$_3$

95

The sharp singlet at 1.71 ppm represents the gem dimethyl group of the bisphenol-A moiety. The complex series of multiplets in the chemical shift range from 6.7 to 7.4 ppm represent the para substitution doublets arising from the aromatic hydrogens ortho to oxygen and gem dimethyl substituents. The doublet at lowest field near 7.83 ppm is due to the hydrogens ortho to the sulfone substituents. The weak band at 1.5 ppm probably represents an impurity in the sample.

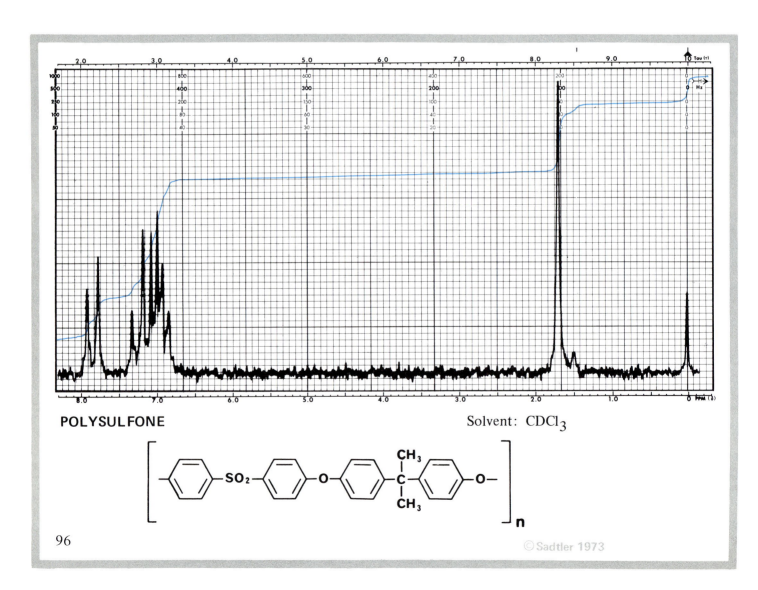

POLYSULFONE

Solvent: $CDCl_3$

Produced by the reaction of sodium polysulfide with dichlorodiethyl formal (Cl-CH$_2$-CH$_2$-O-CH$_2$-O-CH$_2$-CH$_2$-Cl), this polymer displays three sets of absorbance bands in its NMR spectrogram. The methylene groups bonded to sulfur atoms resonate at about 2.95 ppm as a triplet and the corresponding methylene groups bonded to one oxygen atom appear at 3.91 ppm as a triplet. The isolated formal protons (-O-CH$_2$-O-) appear as a sharp singlet at lowest field near 4.75 ppm. This polysulfide has an average molecular weight of about 7500.

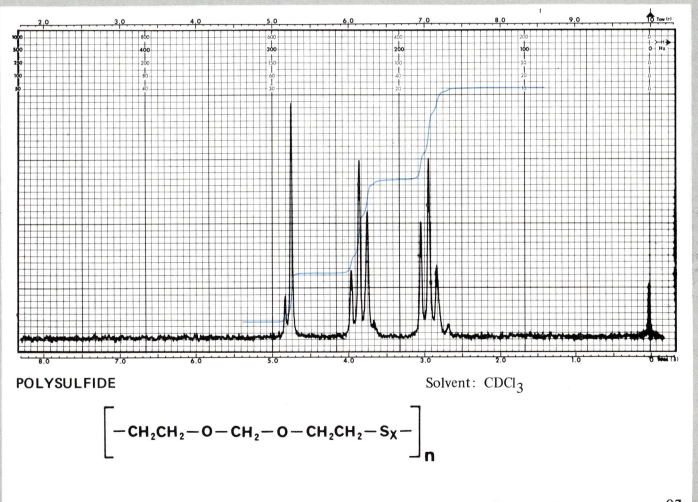

POLYSULFIDE

Solvent: CDCl$_3$

$$\left[-CH_2CH_2-O-CH_2-O-CH_2CH_2-S_X- \right]_n$$

97

This polysulfide which is prepared from sodium polysulfide and dichlorodibutyl formal, produces a spectrogram consisting of four major absorbance bands. The two center CH_2's of the butyl groups resonate at highest field as a complex band centered at about 1.73 ppm. The remaining three bands possess chemical shifts similar to those observed in the previous spectrum: the CH_2 groups bonded to the sulfur atoms resonate at 2.78 ppm, the CH_2's bonded to one oxygen atom appear at 3.52 ppm and the isolated formal CH_2 groups produce a single sharp peak at about 4.62 ppm. A weak triplet centered at about 1.4 ppm represents terminal -SH protons coupled to an adjacent methylene group. The molecular weight of this sample is about 1200.

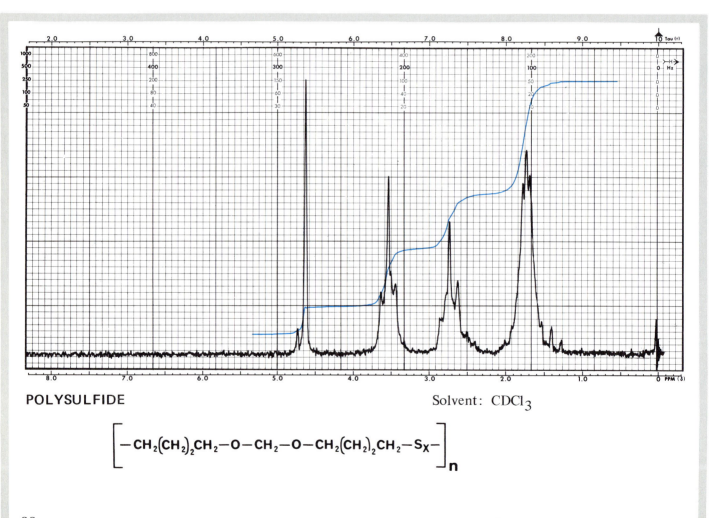

POLYSULFIDE   Solvent: $CDCl_3$

$$\left[-CH_2(CH_2)_2CH_2-O-CH_2-O-CH_2(CH_2)_2CH_2-S_x-\right]_n$$

THE POLYESTERS

The polyester polymers are prepared by the condensation of a difunctional alcohol (a glycol) with a difunctional carboxylic acid. They differ from the analogous nylon polyamides in that certain of the nylons are prepared from a monomer containing both functional groups (an amino acid). A wide variety of different polyesters are utilized in the plastics and coatings industry while polyethylene terephthalate is the polyester used almost exclusively for the preparation of fibers and fabrics.

The acids most commonly used in the formation of polyesters are terephthalic acid, adipic acid, isophthalic acid, and to a lesser extent, sebacic acid and azelaic acid. The commonly employed alcohols are ethylene glycol, diethylene glycol, cyclohexanedimethanol and several of the longer chain aliphatic diols.

Polymerization is usually initiated via an ester interchange reaction in which a prepolymer is formed. When heated under vacuum, the prepolymer loses additional amounts of the glycol leading to a polymer melt possessing the desired molecular weight.

The analysis of the NMR spectra of this group consists of the identification of the two or more monomeric units which comprise the polymer.

NMR Parameters

The characteristic bands which may be present in the NMR spectra of the polyesters are listed below:

1. The methylene groups of the glycol ester linkage ($-CH_2-O-C(=O)-$) normally range in chemical shift from 4.1 ppm to 4.3 ppm for aliphatic acids, and from 4.5 to 4.9 ppm for aromatic acids.

2. The methylene groups of the acid which are adjacent to the ester linkage ($-CH_2-C(=O)-O-$) normally occur in the 2.3 - 2.4 ppm chemical shift range.

3. The four equivalent terephthalic acid protons resonate at about 8.2 ppm as a single peak while the isophthalic acid aromatic hydrogens appear as bands at 7.5, 8.2 and 8.7 ppm.

4. The methylene ether bands of diethylene glycol appear as a distorted triplet near 3.6 ppm.

5. The internal methylene groups of the longer chain aliphatic alcohols or carboxylic acids appear in the 1.4 to 1.7 ppm range.

The last spectrum in this chapter is that of a polycarbonate. The bisphenol-A polycarbonates produce a spectrum consisting of two bands; the four aromatic protons resonate near 7.06 ppm and the two equivalent methyl groups appear at 1.67 ppm.

Nomenclature

There is no generally accepted system of nomenclature for the polyesters. Commercial materials are usually described simply as polyesters. In this text, an attempt has been made to identify all of the monomer units which can be observed in the spectrum and to arrange the compositional names starting with the alcohol(s) followed by the dicarboxylic acid(s). Polyethylene terephthalate is the only polyester which possesses a commonly accepted abbreviation (PET).

The Federal Trade Commission limits the use of the name "polyester fiber" to those containing no less than 85% by weight of a linear polymer composed of terephthalic acid and a diol. The diols most often employed in the production of fibers are ethylene glycol and cyclohexanedimethanol.

Commercial Names

Polyesters

Dacron (1), Dion (48), Hostadur (24), Kodel (50), Koplac (50), Mylar (1), Paraplex (34), Terylene (51), Vycron (52).

Polycarbonates

Carbascar (2), Lexan (46), Merlon (53).

The distorted triplets at 1.68 ppm and 2.37 ppm represent the two methylene absorbance bands of adipic acid. The two CH_2's in the center of the molecule resonate at higher field (1.68 ppm) than the two CH_2 groups which are adjacent to the carbonyl groups. The ethylene glycol protons are observed at 4.28 ppm as a sharp singlet strongly deshielded by the adjacent ester groups. A spectrum of the monomer, adipic acid, is shown on page 229.

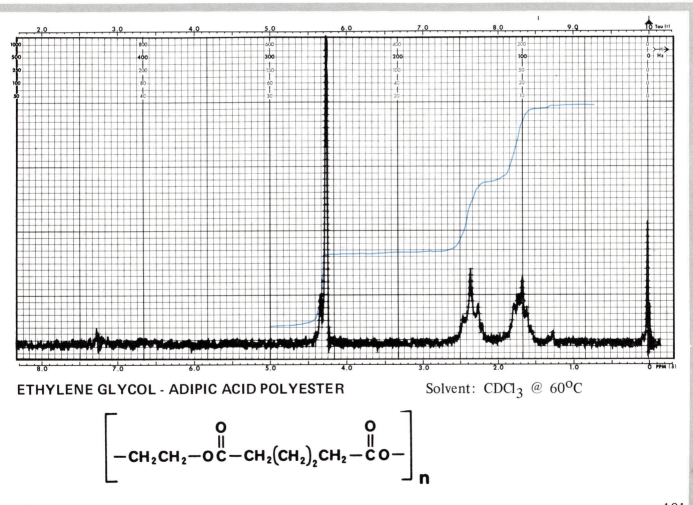

ETHYLENE GLYCOL - ADIPIC ACID POLYESTER   Solvent: $CDCl_3$ @ 60°C

$$\left[-CH_2CH_2-O\overset{O}{\overset{\|}{C}}-CH_2(CH_2)_2CH_2-\overset{O}{\overset{\|}{C}}O- \right]_n$$

101

The absorbance bands arising from the adipic acid protons resonate at 1.63 ppm and 2.32 ppm, virtually identical in position and appearance to those of the previous spectrum. The methylene groups of the diethylene glycol moiety appear as two distorted triplets near 3.6 and 4.15 ppm. The former represents the CH_2 groups bonded to the ether oxygen atom while the latter band arises from the CH_2 groups bonded to the ester oxygen atoms. The weak band at 3.38 ppm is exchangeable and probably represents the absorbance of terminal hydroxyl groups.

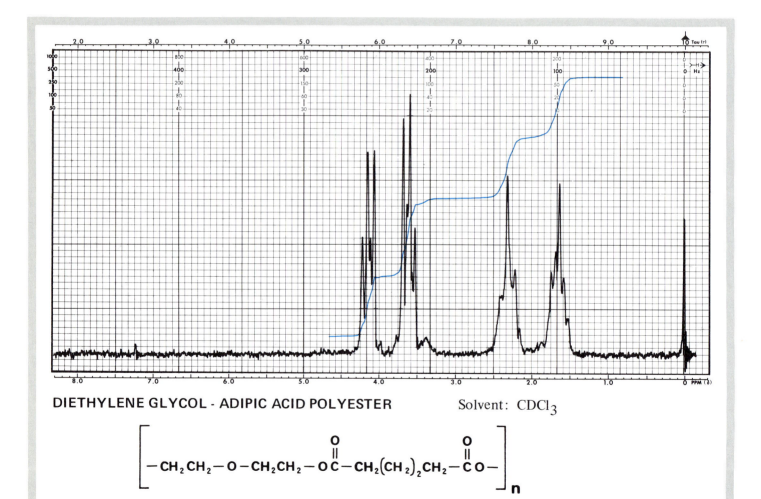

DIETHYLENE GLYCOL - ADIPIC ACID POLYESTER   Solvent: $CDCl_3$

102

This spectrum contains all of the absorption bands of the two previous spectra; ethylene glycol-adipic acid and diethylene glycol-adipic acid polymers. The adipic acid groups of both polyesters possess the same chemical shifts and appear as the two characteristic adipic acid bands at 1.68 ppm and 2.38 ppm. The diethylene glycol methylene groups result in two distorted triplets, one near 3.78 ppm and one near 4.20 ppm. The two equivalent ethylene glycol CH_2 groups appear as a sharp single peak at 4.27 ppm. The peak at 3.02 ppm is due to exchangeable protons and probably represents -OH terminating groups and/or water present in the sample. The ratio of ethylene glycol to diethylene glycol is approximately 2.2 to 1.0.

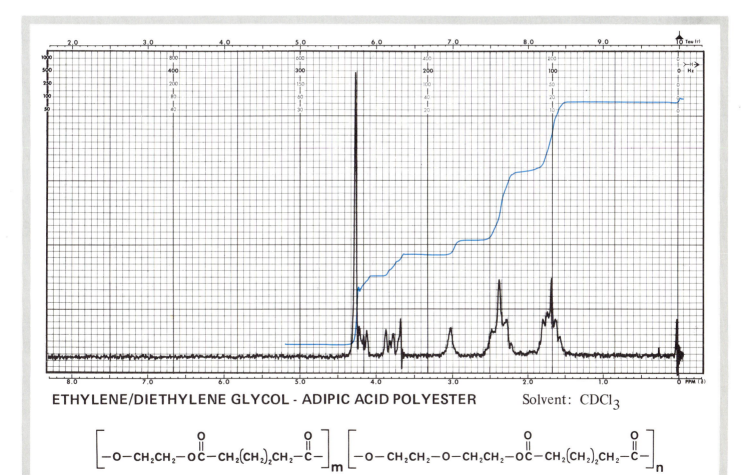

ETHYLENE/DIETHYLENE GLYCOL - ADIPIC ACID POLYESTER   Solvent: $CDCl_3$

103

The complex multiplet in the chemical shift range from 1.6 to 2.0 ppm represents the overlap of the two center methylene groups of adipic acid coinciding in chemical shift with the two center methylene groups of tetramethylene glycol. The adipic acid CH_2's bonded to the ester carbonyl groups resonate at about 2.3 ppm while the tetramethylene glycol CH_2's bonded to the ester oxygen atoms appear near 4.07 ppm. Both of these bands are poorly resolved triplets. The weak triplet near 3.6 ppm and the exchangeable proton band at 4.52 ppm suggest the presence of terminating -CH_2-OH groups.

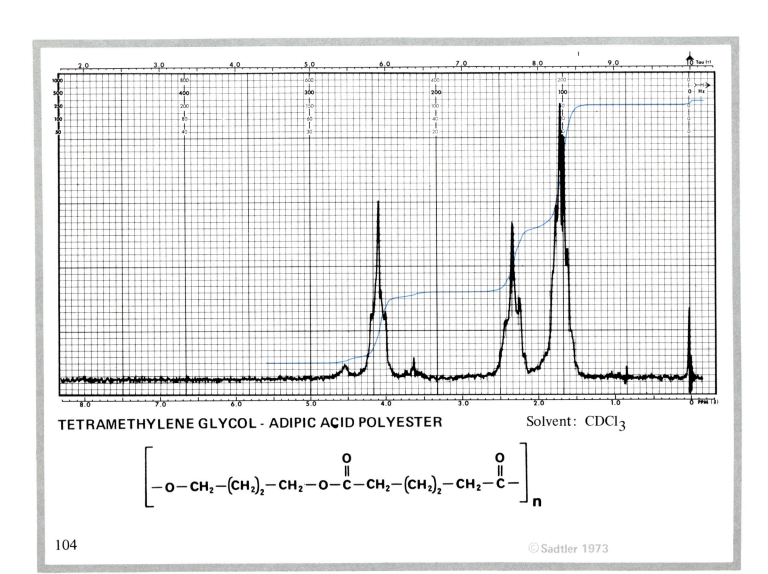

TETRAMETHYLENE GLYCOL - ADIPIC ACID POLYESTER   Solvent: $CDCl_3$

$$\left[-O-CH_2-(CH_2)_2-CH_2-O-\overset{O}{\overset{\|}{C}}-CH_2-(CH_2)_2-CH_2-\overset{O}{\overset{\|}{C}}- \right]_n$$

The adipic acid units are represented by two distorted triplets at 1.62 ppm and 2.29 ppm. The 1,3-butylene glycol units are represented by four different multiplets: a three proton doublet at 1.25 ppm, a two proton quartet at about 1.84 ppm, a triplet at 4.08 ppm and a sextet at 4.95 ppm. The sharp singlet at 2.00 ppm represents terminal acetate groups.

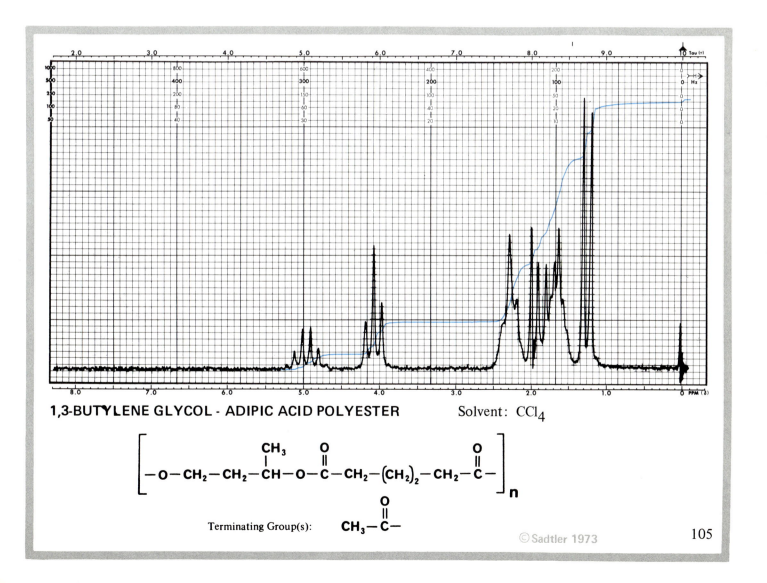

1,3-BUTYLENE GLYCOL - ADIPIC ACID POLYESTER   Solvent: CCl_4

Terminating Group(s): $CH_3 - \overset{O}{\underset{||}{C}} -$

105

POLYESTERS

In comparison with the spectrum of the homopolymer 1,3-butylene glycolpolyadipate on page 105, this spectrum displays additional integration intensity for the bands at 1.65 and 4.08 ppm which represent the absorbance bands of 1,4-tetramethylene glycol. Based upon a comparison of the integration values of the bands at 4.08 ppm and 4.97 ppm, the ratio of tetramethylene glycol units to butylene glycol units is about 1.3 to 1. The singlet at 2.00 ppm represents terminal acetate groups.

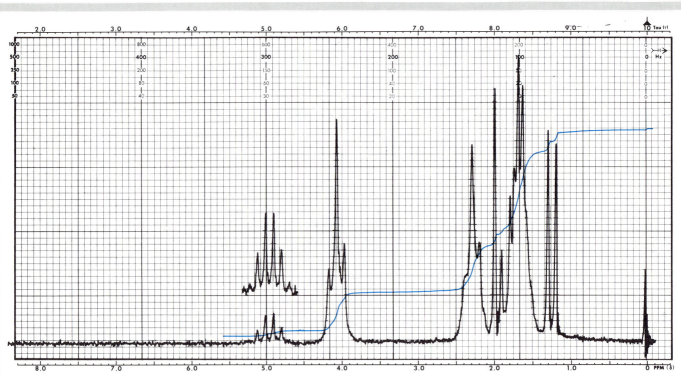

1,4-TETRAMETHYLENE/1,3-BUTYLENE GLYCOL - ADIPIC ACID POLYESTER

Solvent: CCl_4

106

©Sadtler 1973

This spectrum of a diethylene glycol-sebacic acid polyester contains four distinct sets of absorbance bands. The two distorted triplets arising from the diethylene glycol units appear at about 3.7 and 4.23 ppm. The methylene groups of sebacic acid which are adjacent to the carboxylate carbonyl carbon atoms resonate as a broadened triplet at 2.35 ppm. The methylene groups which are beta to the carbonyl carbons appear as a broad multiplet at about 1.63 ppm which overlaps with the remaining CH_2 resonance at highest field. The shape and integration value of this high field absorbance band readily distinguish this polyester from that of the corresponding adipic acid-diethylene glycol polymer on page 102.

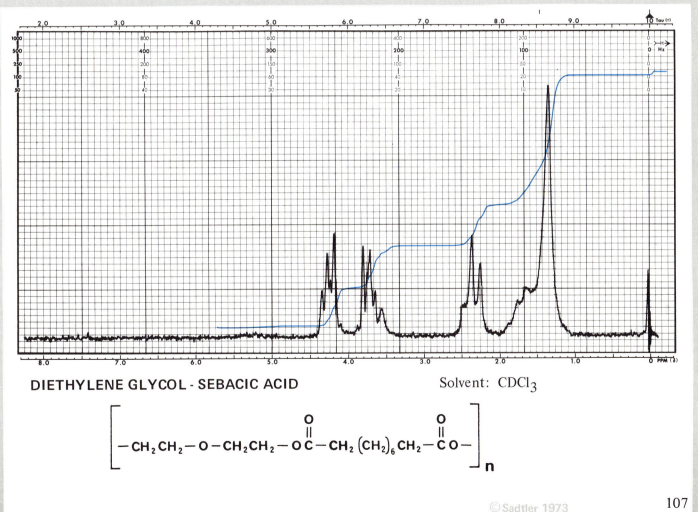

DIETHYLENE GLYCOL - SEBACIC ACID   Solvent: $CDCl_3$

$$\left[-CH_2CH_2-O-CH_2CH_2-O\overset{\overset{\displaystyle O}{\|}}{C}-CH_2(CH_2)_6CH_2-\overset{\overset{\displaystyle O}{\|}}{C}O- \right]_n$$

107

The four equivalent aromatic protons of each polyethylene terephthalate unit are strongly deshielded by the adjacent ester carbonyl groups and as a result appear as a sharp singlet at very low field, near 8.2 ppm. The four equivalent methylene protons of each ethylene glycol moiety resonate near 4.88 ppm as a single narrow peak. This polymer is widely utilized as a synthetic fiber and is produced by a two stage ester exchange reaction between dimethyl terephthalate and ethylene glycol.

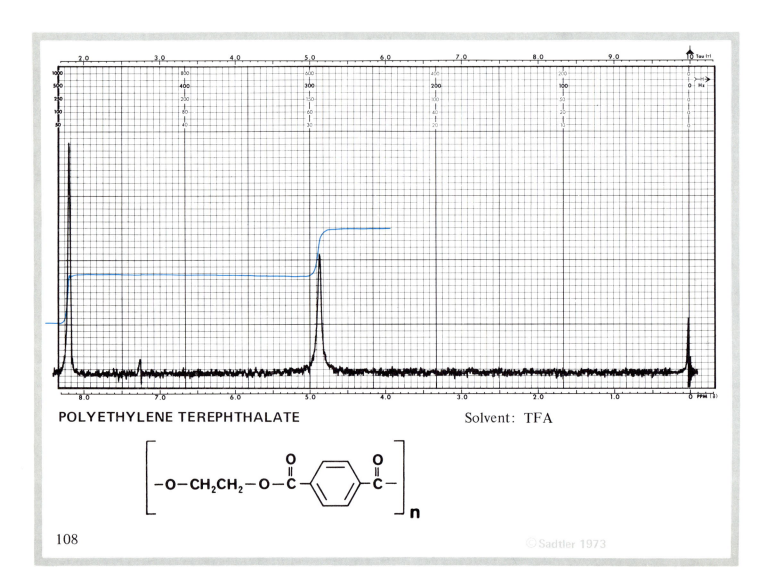

POLYETHYLENE TEREPHTHALATE   Solvent: TFA

108

The two methylenes which are bonded to the oxygen atoms of the ester groups appear as a broad multiplet centered at about 4.45 ppm while the cyclohexane ring hydrogens are represented by a series of poorly resolved multiplets in the chemical shift range from 1.0 to 2.5 ppm. As in the previous spectrum, the four equivalent aromatic protons resonate as a sharp singlet near 8.27 ppm. The complexity of the methylene band at 4.45 ppm suggests that the polymer contains both the cis- and trans- cyclohexanedimethanol isomers.

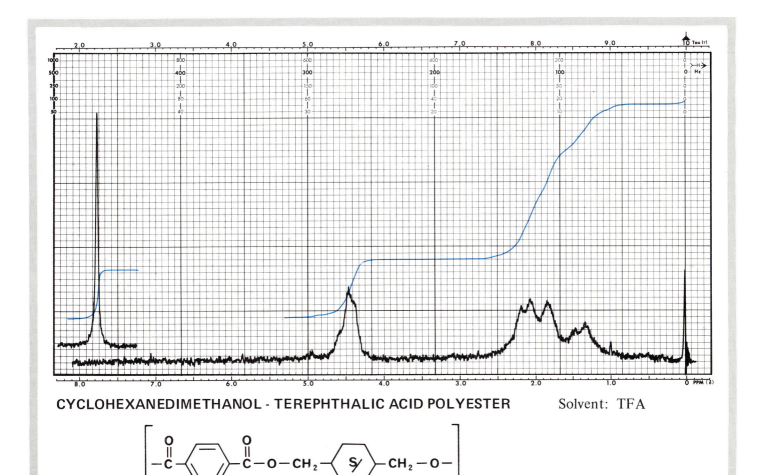

CYCLOHEXANEDIMETHANOL - TEREPHTHALIC ACID POLYESTER   Solvent: TFA

109

POLYESTERS

The propylene glycol methyl groups resonate near 1.45 ppm as a complex multiplet due to the overlap of methyl doublets in differing polymer environments. The glycol methylene groups appear at 4.98 ppm and the methines at about 5.5 ppm. The isophthalic acid protons occur as three multiplets, a triplet at 7.51 ppm (H-5), a two hydrogen doublet at 8.22 ppm (H-4 and H-6) and a broadened dimeta triplet at 8.68 ppm (H-2). The single peak at 6.9 ppm represents the olefinic protons of the fumaric acid units. The ratio of isophthalic acid to fumaric acid is approximately 2.3 to 1.

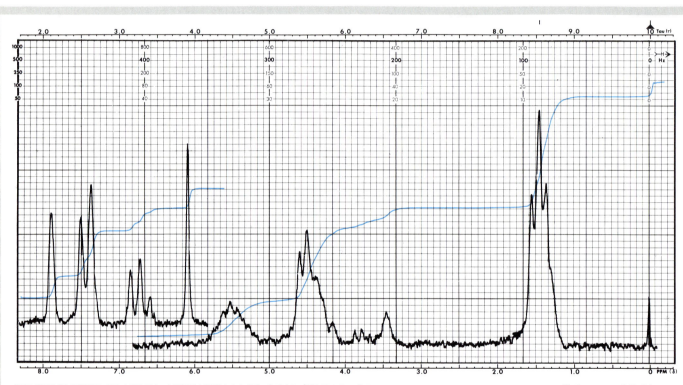

PROPYLENE GLYCOL-ISOPHTHALIC ACID/FUMARIC ACID POLYESTER   Solvent: $CDCl_3$

110

This polyester has been prepared from tetramethylene glycol, azelaic acid, isophthalic acid and terephthalic acid. The azelaic acid bands appear at 1.3 and 2.3 ppm. The CH_2 groups bonded to the azelaic acid oxygen atoms appear as a weak triplet at 4.12 ppm, while those bonded to the aromatic acid oxygen atoms appear at 4.42 ppm. The sharp singlet at 8.05 ppm represents the terephthalic acid protons. The isophthalic acid aromatic hydrogens occur as a complex series of multiplets in the range from 7.3 to 8.7 ppm. The mole ratio of acids present in the sample is approximately 6:3:2 (isophthalic, terephthalic, azelaic).

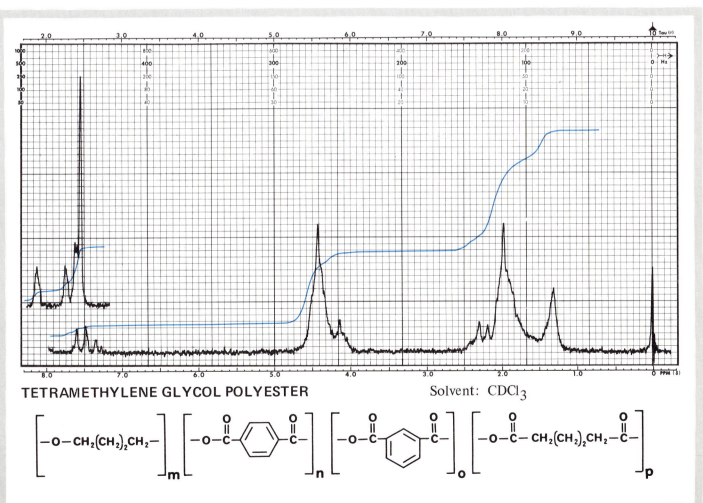

TETRAMETHYLENE GLYCOL POLYESTER   Solvent: $CDCl_3$

111

POLYESTERS

The absorption bands arising from tetramethylene glycol appear at 1.95, 4.13 and 4.42 ppm. The multiplets at 7.49, 8.19 and 8.62 ppm represent the isophthalic acid hydrogens while the singlet at 8.06 ppm represents terephthalic acid. Weak bands at 1.26, 1.51 and 2.83 ppm appear to arise from the presence of an aliphatic acid group which cannot be characterized because of its low concentration in the sample.

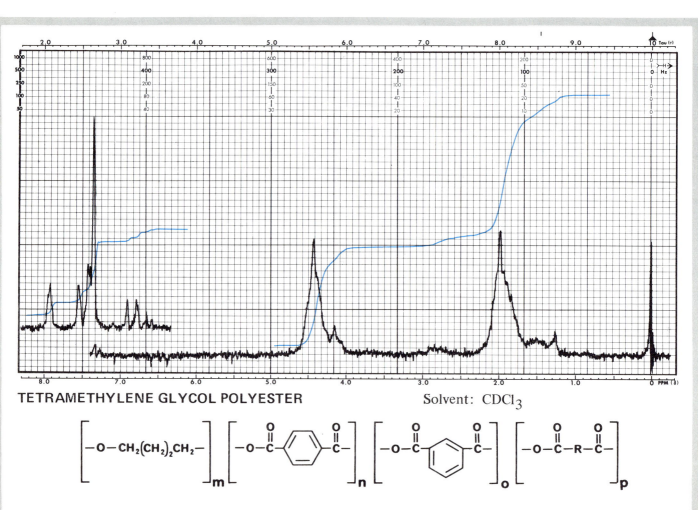

TETRAMETHYLENE GLYCOL POLYESTER

Solvent: $CDCl_3$

© Sadtler 1973

The dominant absorption bands of this spectrum which appear in the chemical shift range from 5.1 to 7.5 ppm represent the monomeric styrene which is present in the sample (compare with the spectrum on page 196). The alcoholic moiety, which appears to be 1,2-propylene glycol, is represented by broad, poorly resolved bands at 1.34, 4.38 and 5.40 ppm. Two different dibasic acids were used in the preparation of the polyester which are represented by a single peak at 6.9 ppm (fumaric acid) and a complex multiplet centered at 7.68 ppm which may be due to the presence of phthalic acid.

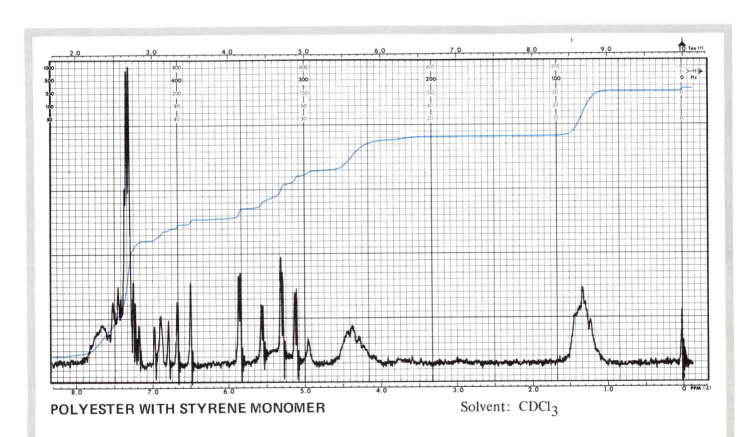

POLYESTER WITH STYRENE MONOMER   Solvent: CDCl$_3$

COMPLEX STRUCTURE

In CDCl$_3$ solutions, bisphenol-A polycarbonates produce a simple spectrum consisting of two single peaks, one at 7.06 ppm representing eight aromatic protons and one at 1.67 ppm representing the six protons of the gem dimethyl bridging group. The aromatic protons provide a single peak in CDCl$_3$ solution but display a typical para substitution pattern in trifluoroacetic acid (TFA) consisting of two doublets.

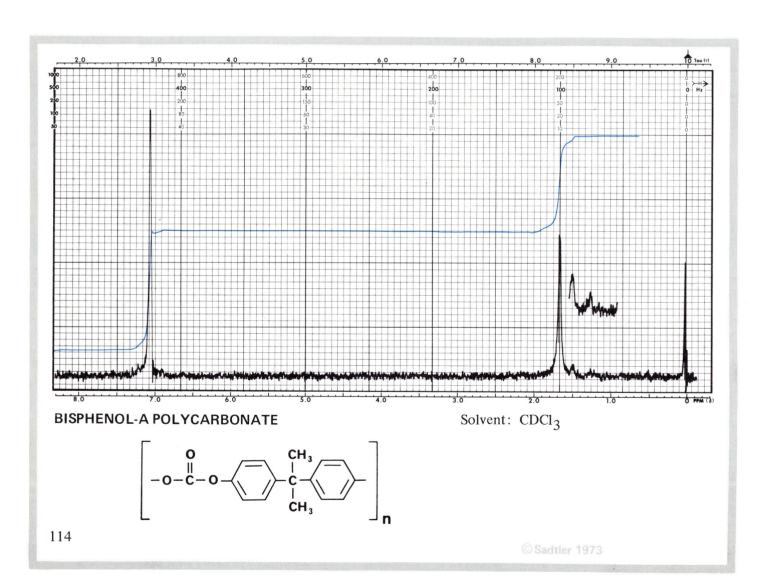

BISPHENOL-A POLYCARBONATE   Solvent: CDCl$_3$

THE POLYAMIDES

The term Nylon has become almost synonymous with the class of polymers which are chemically designated as the polyamides. The nylons are of two chemical varieties; those derived from diamines and dicarboxylic acids, and those formed from amino acids or their cyclic precursors, the lactams.

$$[-NH-(CH_2)_x-NH-(C=O)-(CH_2)_y-(C=O)-]_n \qquad \text{AABB type} \qquad \text{(nylon 6,6)}$$

$$[-NH-(CH_2)_x-(C=O)-]_n \qquad \text{AB type} \qquad \text{(nylon 6)}$$

The numerical designations of the different nylon types is based upon the number of carbon atoms in each repeating polymer unit. If the polymer contains only one monomeric group which contains eleven carbons, it would be named nylon 11. If the nylon was prepared from a diamine containing six carbons and a dicarboxylic acid containing ten carbons, it would be named nylon 6,10.

The term polyamide is also applied to commercial resins prepared by the condensation of ethylene diamine with a fatty polycarboxylic acid. The physical properties and applications of these resins are quite different from those of the nylons and these compounds will be discussed separately later on in this chapter.

The nylons are commercially prepared by three main routes. The first and most important process involves the formation of a salt by stoichiometric combination of a suitable diamine and dicarboxylic acid. The salt is then thermally polymerized, with water being present as both a catalyst and a by-product of the reaction. Less commonly employed, but useful for certain specific nylons, is the interfacial polymerization of diacyl chlorides with diamines. The third process utilizes either lactams (cyclic amides) or amino acids, and like the first reaction, requires both high temperatures and a catalyst such as water. In this case, however, no additional water is liberated as a by-product.

As with the polyethers and polyesters, the nylons are made up of relatively small hydrocarbon fragments alternating with a functional group, in this case the amide linkage. Thus the NMR spectra of these polymers contain a combination of the hydrocarbon segments derived from the monomers.

NMR Parameters

The NMR spectra of the polyamides can for the most part be described by three or four characteristic bands:

1. The methylene groups attached to the amide nitrogen atoms ($-CH_2-NH-C(=O)-$) which are usually seen between 3.5 - 3.6 ppm depending upon the solvent employed.

2. The methylene groups attached to the amide carbonyl carbon atom $(-CH_2-C(=O)-NH-)$ which resonates in the 2.65 - 2.75 ppm region.

3. The methylene groups which are beta to or farther removed from the amide nitrogen and carbonyl carbon atoms. These bands are observed in the 1.4 - 1.9 ppm range with the methylene groups beta to the nitrogen atoms usually appearing at lower field (1.7 - 1.9 ppm).

The ethylene diamine-fatty acid condensates produce spectra which are readily distinguishable from those of the nylons. Their spectra contain terminal methyl groups at 0.9 ppm, a large number of chain methylene groups which resonate at about 1.3 ppm and the two ethylene diamine methylene groups which resonate as a broad single peak at about 3.4 ppm.

All of the chemical shifts described above are solvent dependent. The two solvents which are most effective in solubilizing the nylons are trifluoroacetic acid and formic acid. The former (TFA), which is a stronger acid, will generally produce chemical shifts which are downfield from those observed in formic acid. This observation should be kept in mind when comparing unknown spectra to those recorded in the literature.

Because the bands discussed above are all well separated in chemical shift, accurate integration values are usually obtained which can be very helpful in determining the molecular structure of a specific homopolymer.

The table presented below lists the expected integration ratios for most commercial and experimental nylons. In those cases where two different compounds produce the same integration ratio, the identity of the unknown can often be determined by comparing the high field absorbance pattern (1.4 - 1.9 ppm) with that of a known reference compound.

Nylon Type	Integration ratio		
	CH_2-NH	$CH_2-C(=O)$	$(CH_2)_n$
Nylon 4	1	1	1
Nylon 6	1	1	3
Nylon 6,6	1	1	3
Nylon 6,10	1	1	5
Nylon 8	1	1	5
Nylon 11	1	1	8
Nylon 12	1	1	9

Commercial Names

Nylon 6
Capran (54), Caprolan (55), Durethan (43), polycaprolactam, polycaproamide.

Nylon 6,6
Nylatron GS (56), Zytel (1), polyhexamethylene adipamide.

Nylon 6,10
Polyhexamethylene sebacamide.

Nylon 11
Rilsan B (57), polyundecanoamide.

Nylon 12
Rilsan A (57), Vestamid (58), polylauroamide, polylaurolactam.

Fatty acid polyamides
Emerez (59), Versalon (60), Versamid (60).

The band at 3.58 ppm represents the resonance of the methylene groups which are bonded to nitrogen atoms while the band at 2.75 ppm is due to the CH_2's which are bonded to the carbonyl groups. The remaining methylene protons overlap to form a complex band at high field in the chemical shift range from 1.1 to 2.0 ppm. The appearance of this high field band is characteristic to a nylon 6 and is useful in distinguishing a spectrum of this material from that of a nylon 6, 6 which would possess the same integration values (see page 119).

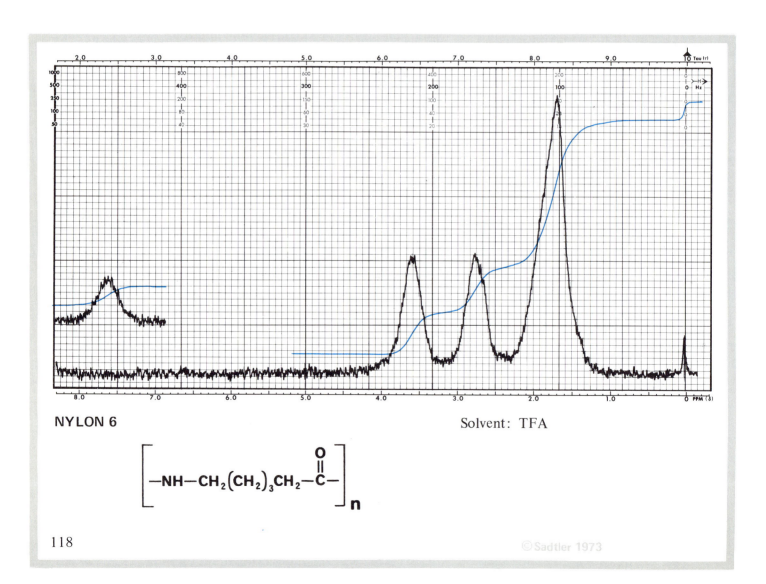

NYLON 6   Solvent: TFA

$$\left[-NH-CH_2\left(CH_2\right)_3 CH_2-\overset{\overset{\displaystyle O}{\|}}{C}- \right]_n$$

POLYAMIDES

The band centered at 1.5 ppm is due to the center methylene groups of the hexanediamine moiety while the band near 1.9 ppm represents the center methylene groups of the adipic acid segments. The broadened triplet at 2.7 ppm arises from the methylenes bonded to the carbonyl groups and the band at 3.5 ppm arises from those CH_2 groups bonded to the nitrogen atoms. At 8.3 ppm is observed a broad band due to the secondary amide protons in the polymer.

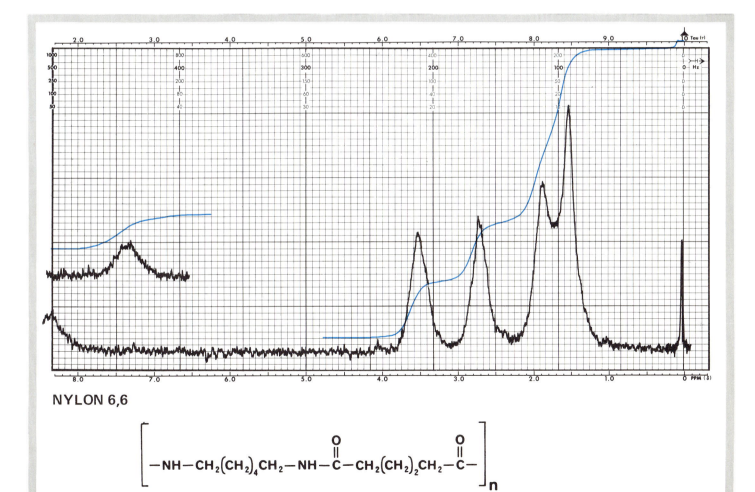

NYLON 6,6

119

The methylene groups in the middle of the chain form a complex band at high field in the chemical shift range from 1.2 to 2.1 ppm. Its appearance and relative integration value allows one to distinguish it from other types of nylon. The CH_2 group bonded to the amide nitrogen group resonates at about 3.5 ppm and the CH_2 group bonded to the carbonyl appears at about 2.62 ppm.

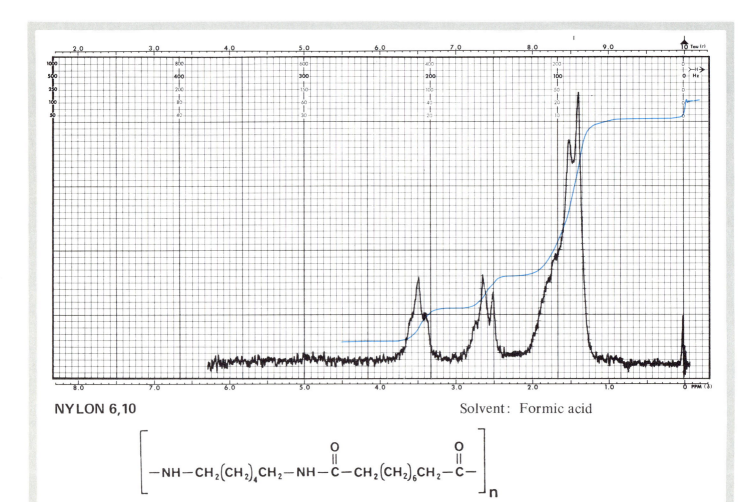

NYLON 6,10   Solvent: Formic acid

$$\left[-NH-CH_2(CH_2)_4CH_2-NH-\overset{\overset{\displaystyle O}{\|}}{C}-CH_2(CH_2)_6CH_2-\overset{\overset{\displaystyle O}{\|}}{C}- \right]_n$$

This NMR spectrogram is that of a nylon molding resin based upon nylon 6,10. The chemical shifts and band shapes are very similar to those of the preceding spectrum. As in the spectra of the other nylons, the band at about 3.36 ppm arises from the methylene groups bonded to the amide nitrogen atoms, while the band near 2.42 ppm represents the CH_2's bonded to the carbonyl groups. The remaining ten methylenes form a complex higher order envelope of absorbance in the chemical shift range from 1.2 to 2.0 ppm. The amide NH protons overlap with the formic acid protons at low field (6.5 to 8.5 ppm).

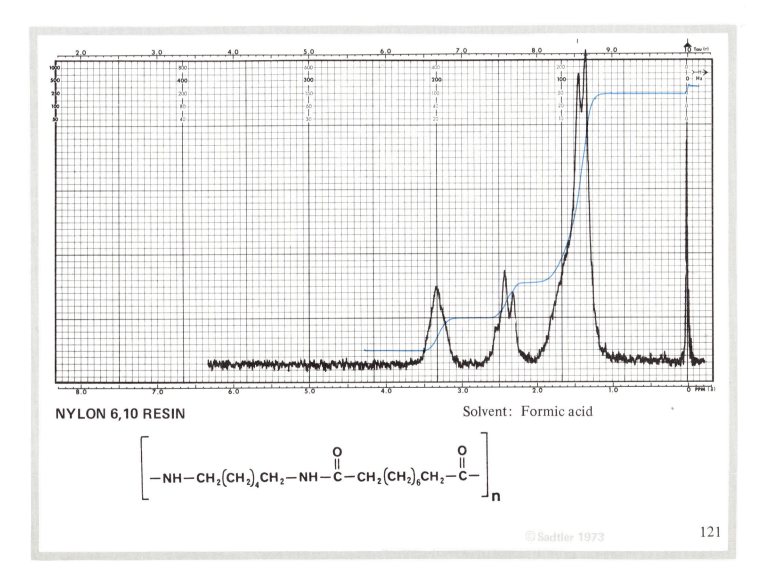

NYLON 6,10 RESIN   Solvent: Formic acid

$$\left[-NH-CH_2(CH_2)_4 CH_2 - NH - \overset{\overset{O}{\|}}{C} - CH_2(CH_2)_6 CH_2 - \overset{\overset{O}{\|}}{C} - \right]_n$$

121

Because nylon 11 absorbs less water than nylon 6 or nylon 6,6 it has found many applications in the field of electrical components. Its spectrum is characterized by a strong narrow peak at high field representing the eight methylene groups in the center of the polymer unit. The methylene groups bonded to the amide carbonyl and NH groups appear at 2.75 and 3.6 ppm respectively. The basic integration ratio of 1:1:8 distinguishes a spectrum of nylon 11 from that of nylon 12 which is similar in appearance (see page 123).

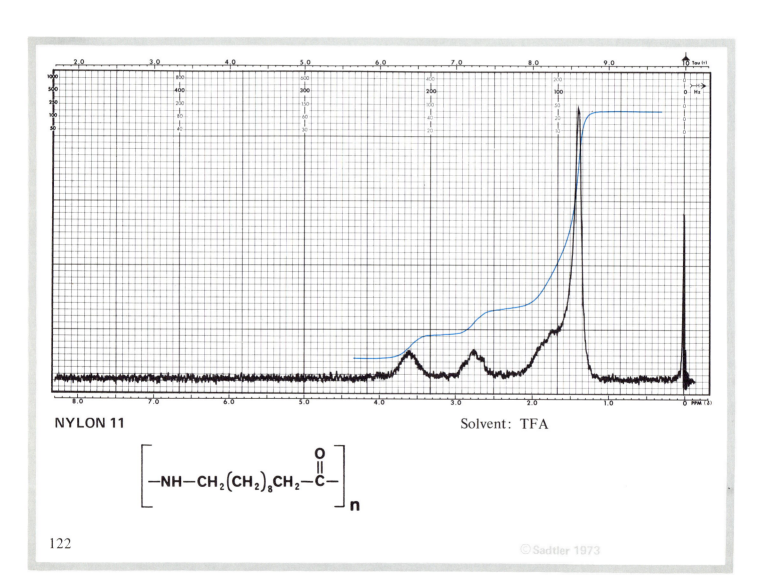

NYLON 11   Solvent: TFA

$$\left[-NH-CH_2\left(CH_2\right)_8 CH_2-\overset{\displaystyle O}{\overset{\displaystyle \|}{C}}- \right]_n$$

Nylon 12 possesses lower specific gravity, melting point and moisture absorption than the other nylons. Its NMR spectrogram is very similar to that of nylon 11 but possesses a basic integration ratio of 1:1:9. The very weak absorption band due to the amide protons is usually not observed except at very high sample concentrations which can rarely be obtained. The methylene groups bonded to the amide carbonyl and nitrogen atoms appear as broadened triplets at 2.7 and 3.6 ppm. The remaining CH_2's resonate at higher field in the chemical shift range from 1.1 to 2.1 ppm.

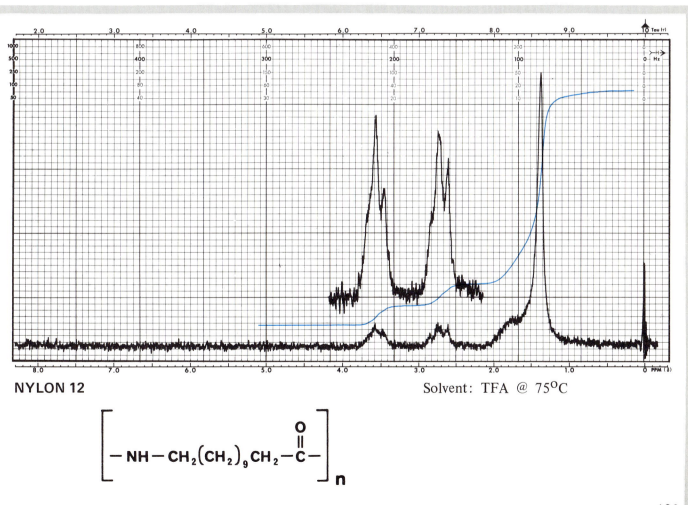

NYLON 12 Solvent: TFA @ 75°C

123

This spectrum of a nylon 6/6,6 polymer is very similar in appearance to the spectra of the two homo-polymers. The shape and position of the maxima of the high field absorbance bands appears to be inter-mediate between that of the nylon 6 spectrum and that of the nylon 6,6 spectrum. Based upon the relative peak heights of the band at 1.74 ppm and the band at 1.52 ppm this sample appears to contain the two nylon homopolymers at a ratio of about 1:1.

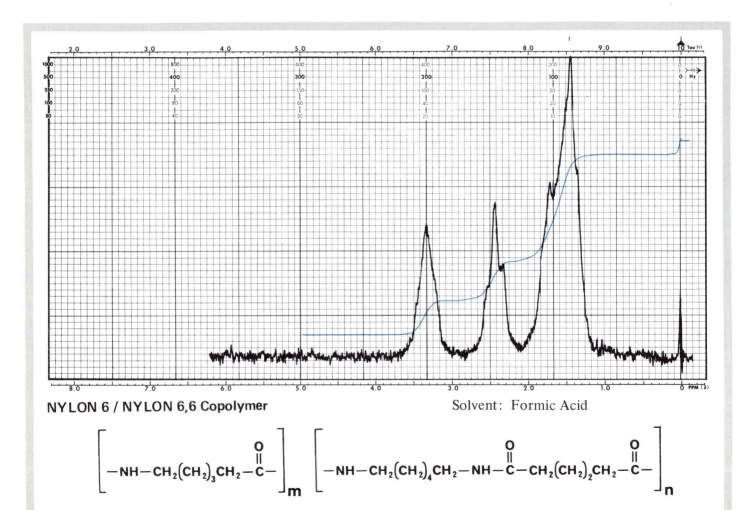

NYLON 6 / NYLON 6,6 Copolymer

Solvent: Formic Acid

Polyamide resins of this type are characterized by the appearance of a saturated fatty acid moiety at high field (0.6 to 2.8 ppm), a broadened single peak near 3.4 ppm representing the ethylene diamine protons (N-CH$_2$CH$_2$-N) and a broad, exchangeable proton band at low field (6 to 8 ppm) representing the amide NH protons. These resins are usually prepared from dimer acid (see page 239) and are used in the manufacture of printing inks, paints and adhesives.

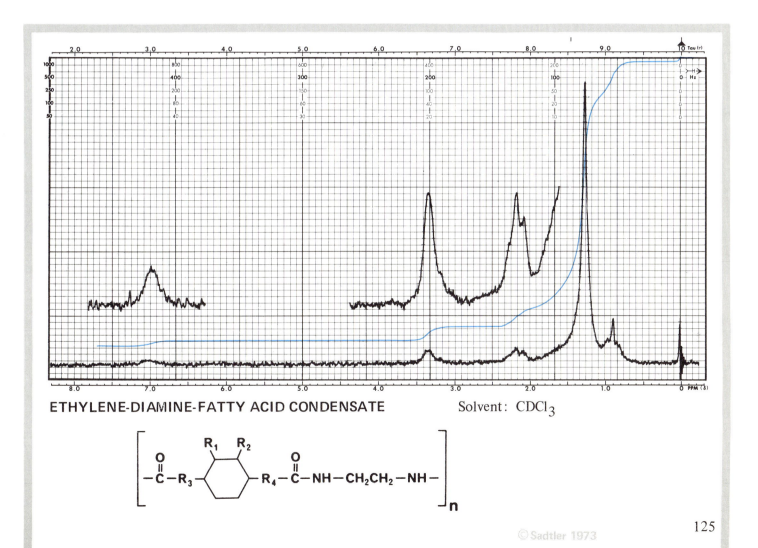

ETHYLENE-DIAMINE-FATTY ACID CONDENSATE   Solvent: CDCl$_3$

125

This polyamide spectrum is of the same type as that scanned on the previous page. The band shapes and position are quite similar. The integration values of the fatty acid chain is significantly lower indicating that a shorter dimer acid chain was used in the preparation of this resin. The broad single peak near 3.36 ppm characterizes this spectrum as that of a fatty acid-ethylene diamine condensate.

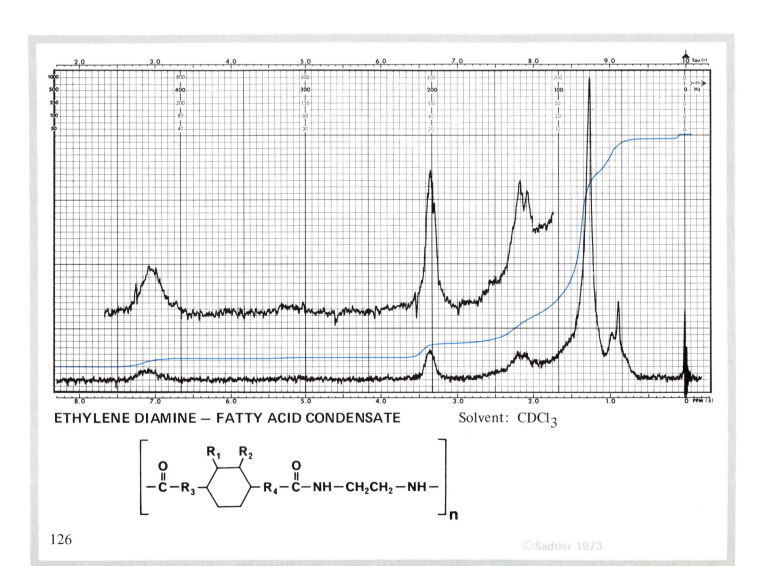

ETHYLENE DIAMINE – FATTY ACID CONDENSATE   Solvent: CDCl$_3$

THE POLYURETHANES

The term polyurethane implies that the polymer consists of repeating units containing the substituted carbamic acid function (R–NH–C(=O)–O–R). Such a molecule could theoretically be formed by the reaction of a diol with a diisocyanate, as shown below.

$$HO-R_1-OH \quad + \quad O=C=N-R_2-N=C=O \quad \longrightarrow$$

$$[-O-R_1-O-C(=O)-NH-R_2-NH-C(=O)-]_n$$

Most commercial polyurethanes, however, contain only a few widely spaced urethane linkages. This is due to the fact that polyethers and polyesters possessing terminal hydroxyl groups are used instead of simple monomeric glycols. The polyethers and polyesters which are employed are of relatively low molecular weight (500 - 5000) and these short polymer chains are then joined by a urethane linkage producing a molecule with a greatly extended chain length.

$$H-(-O-R_1-)_m-OH \quad + \quad O=C=N-R_2-N=C=O \quad \longrightarrow$$

$$[(-O-R_1-)_m-O(C=O)-NH-R_2-NH-C(=O)-]_n$$

The polyols which are most commonly employed in the synthesis of polyurethanes are polyethylene glycol, polypropylene glycol and block copolymers of these two prepolymers. In addition, polyfunctional materials are also prepared from triols, tetrols and hexols which have been polymerized with polypropylene oxide to give prepolymers which when reacted with diisocyanates lead to rigid crosslinked urethanes.

The polyester precursors are usually made from adipic acid or phthalic acid in combination with either ethylene or propylene glycol. To insure the presence of terminal hydroxyl groups, these prepolymers are prepared using an excess of the glycol.

The diisocyanates which are most often employed are aromatic rather than aliphatic, with tolylene diisocyanate (TDI) and diphenylmethane diisocyanate (MDI) being the most popular of these agents. The table on page 128 shows the names, structures and abbreviations for these and other important diisocyanates.

DADI	DIANISIDINE DIISOCYANATE	
HDI	1,6-HEXAMETHYLENE DIISOCYANATE	$O=C=N-CH_2(CH_2)_4-CH_2-N=C=O$
MDI	4,4'-DIPHENYLMETHANE DIISOCYANATE	
NDI	1,5-NAPHTHALENE DIISOCYANATE	
PDI	PHENYLENE DIISOCYANATE	
TODI	TOLIDINE DIISOCYANATE	
TDI	2,4/2,6-TOLYLENE DIISOCYANATE	and/or
XDI	XENYLENE DIISOCYANATE	

A second method of achieving a crosslinked polymer utilizes an excess of isocyanate which under the proper catalysis forms the allophanate linkage shown below.

$$[-O-R_1-O-C(=O)-NH-R_2-NH-C(=O)-]_n \quad + \quad O=C=N-R_3-N=C=O \longrightarrow$$

$$[-O-R_1-O-C(=O)-N-R_2-NH-C(=O)-]_n$$
$$|$$
$$O=C-NH-R_3-NH-C(=O)-$$

NMR Parameters

The NMR spectra of the urethanes will, to a large extent, resemble those of the polyethers and polyesters from which they were formed. The polyurethanes are distinguished from their prepolymers by the appearance in their spectra of fairly weak bands arising from the diisocyanates used to form the urethane linkages. Since most of the diisocyanates are aromatic, their absorbance bands are not obscured by those originating from the polyesters and polyethers.

The aromatic protons of MDI (diphenylmethane diisocyanate) appear as a singlet at 7.21 ppm in TFA but as a narrow multiplet near 7.03 ppm in $CDCl_3$ solutions. The methylene bridging groups resonate at 3.88 ppm in $CDCl_3$ and at 3.95 in TFA solutions.

The aromatic pattern produced by TDI (tolylene diisocyanate) is more complex, since the material used is often a mixture of the 2,4- and 2,6- isomers. TDI produces three aromatic multiplets which appear at about 7.05, 7.27 and 7.67 ppm. The methyl groups attached to the rings resonate at about 2.2 ppm. The methyl groups attached to the rings resonate at about 2.2 ppm. These chemical shifts were obtained from samples examined in CCl_4 solutions. The NMR resonances of the polyester and polyether protons occur at approximately the same chemical shifts as they do in normal ether and ester polymers, i.e.

Internal methylenes from adipic acid	1.87 ppm
Methylenes adjacent to carbonyl groups	2.53 ppm
Methylenes adjacent to oxygen groups	4.1 - 4.3 ppm
Ethylene glycol methylene groups	4.5 ppm

Solubility

Most high molecular weight or crosslinked polyurethanes are insufficiently soluble for the preparation of NMR spectra. Lower molecular weight varieties may be soluble in organic solvents such as CCl_4, $CDCl_3$, DMSO-d6 or TFA.

Applications

The urethanes and other isocyanate-derived polymers and resins have found a wide variety of commercial applications as millable rubbers, elastomeric thread, flexible and rigid foams, adhesives and coatings.

Commercial Products

The numbers in parentheses refer to the list of manufacturers on page 275 to whom these copyright names belong.

Carbamac (61), Castethane (62), Densite (63), Desmoflex (43), Dyalon (64), Estane (23), Everlon (65), Lycra (1), Unithane (49).

This low polymer is similar in appearance and chemical shifts to its monomer, diphenyl methane diiso-cyanate. The CH_2 groups situated between the aromatic rings resonate as a broadened singlet at about 3.88 ppm. The para substituted aromatic rings produce a higher order pattern which is centered at about 7.03 ppm. The NH protons of the urea linkage were not observed but probably resonate as a very broad, weak band in the offset range below 8.0 ppm.

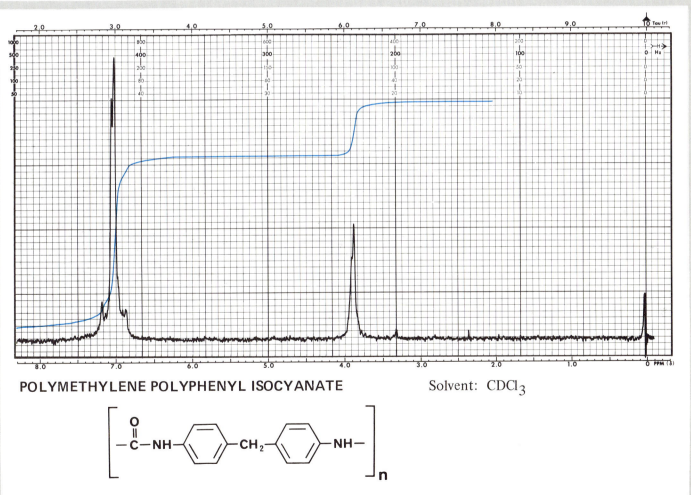

POLYMETHYLENE POLYPHENYL ISOCYANATE   Solvent: $CDCl_3$

131

POLYURETHANES

This elastomeric polyurethane sample is composed of an ethylene glycol-adipic acid polyester modified by diphenyl methane diisocyanate. The distorted adipic acid triplets are observed at 1.75 ppm and 2.55 ppm. The two equivalent ethylene glycol units appear as a single sharp peak at 4.49 ppm. The methylene group of MDI resonates as a weak broadened band at 3.95 ppm. The aromatic protons appear as a single peak with a chemical shift of about 7.21 ppm. Additional absorption is observed on the integration scan in the chemical shift range from 6.8 to 8.2 ppm which represents the urethane amide proton absorbance. The ratio of ethylene glycol to adipic acid to MDI is about 1.6 to 1.5 to 1.

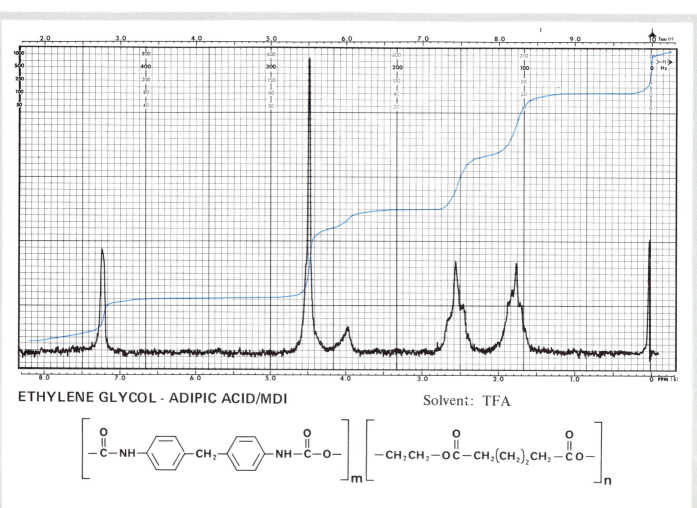

ETHYLENE GLYCOL - ADIPIC ACID/MDI   Solvent: TFA

This polyurethane elastomer was formed by the reaction of diphenyl methane diisocyanate (MDI) with an adipic acid-tetramethylene glycol polyester. The central methylene groups of both adipic acid and tetramethylene glycol overlap to form the broad band centered at about 1.87 ppm. The methylene groups adjacent to the carbonyl carbons resonate at 2.53 ppm and those adjacent to the oxygen atoms appear at 4.3 ppm. The weak band at 3.9 ppm represents the diisocyanate methylene groups. The aromatic hydrogens in TFA solution are accidentally equivalent and resonate as a single peak at 7.17 ppm. The mole ratio of diol to diacid to diisocyanate is approximately 4 to 3 to 1.

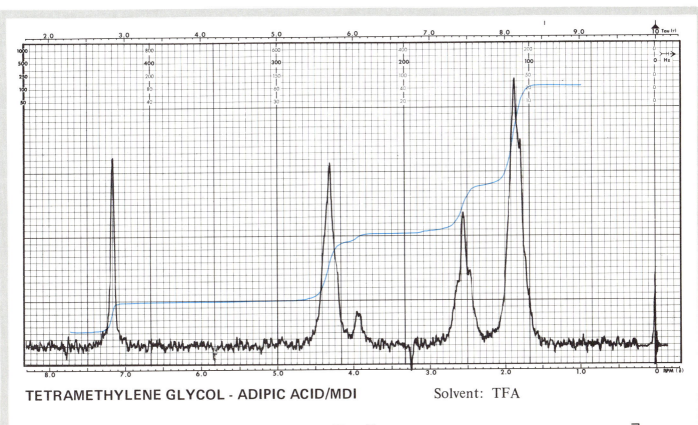

TETRAMETHYLENE GLYCOL - ADIPIC ACID/MDI   Solvent: TFA

133

POLYURETHANES

This polyurethane was prepared from an adipic acid polyester containing two different diol units; ethylene glycol and tetramethylene glycol. The band centered at 1.67 ppm represents the resonance of the center methylene groups of both adipic acid and tetramethylene glycol. The adipic acid CH_2's which are adjacent to the carbonyl groups resonate as a broadened triplet at 2.34 ppm. The tetramethylene glycol CH_2's which are adjacent to the carboxylate oxygen atoms appear at 4.12 ppm. The sharp singlet at 4.3 ppm arises from the two equivalent methylene groups of ethylene glycol. The bands arising from diphenyl methane diiso- appear at 3.89, 7.1 and 7.33 ppm. The urethane amide protons were not observed.

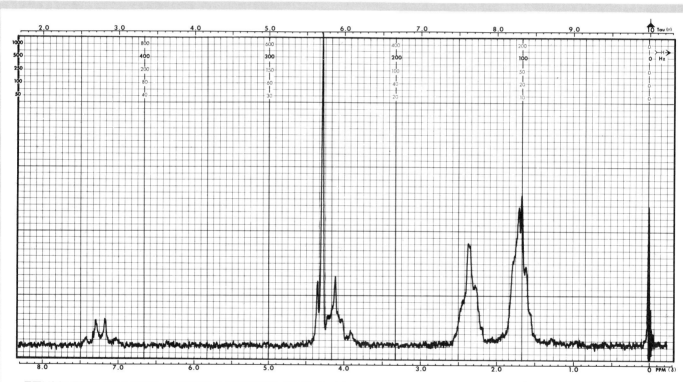

ETHYLENE GLYCOL / TETRAMETHYLENE GLYCOL - ADIPIC ACID/MDI   Solvent: $CDCl_3$

COMPLEX STRUCTURE

134

The bands at high field – 1.7, 2.7 and part of the band at 3.9 ppm – represent the aliphatic urethane unit prepared by the reaction of 1,6-hexamethylene diisocyanate with tetramethylene glycol. The remaining bands in the spectrum indicate the presence of para substituted diphenyl methane units. This aromatic component is represented by part of the band at 3.9 ppm, and the bands at 7.05, 7.38 ppm and 11.42 ppm. The last band which appears in the offset range was found to be exchangeable.

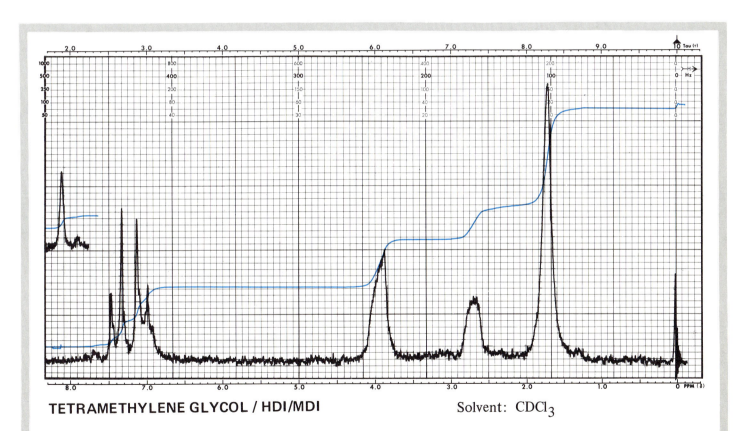

TETRAMETHYLENE GLYCOL / HDI/MDI

Solvent: $CDCl_3$

COMPLEX STRUCTURE

POLYURETHANES

This urethane sample was prepared by reacting the polyether of tetramethylene glycol with tolylene diisocyanate (TDI). The two familiar tetramethylene glycol bands appear at 1.61 ppm and 3.61 ppm. The methylene groups adjacent to the terminal hydroxyl groups have been shifted to lower field upon formation of the urethane linkages with the diisocyanate.

$$-CH_2-OH \quad + \quad O=C=N-\emptyset \quad \longrightarrow \quad -CH_2-O-C(=O)-NH-\emptyset$$

These deshielded methylene groups resonate as a broadened triplet at about 4.15 ppm. The diisocyanate methyl groups appear as two closely spaced peaks near 2.25 ppm while the aromatic hydrogens produce a complex series of peaks in the chemical shift range from 6.7 to 7.8 ppm. The tolylene diisocyanate used to prepare this urethane polymer was probably one of the isomeric mixtures which are presented on p. 224, 226.

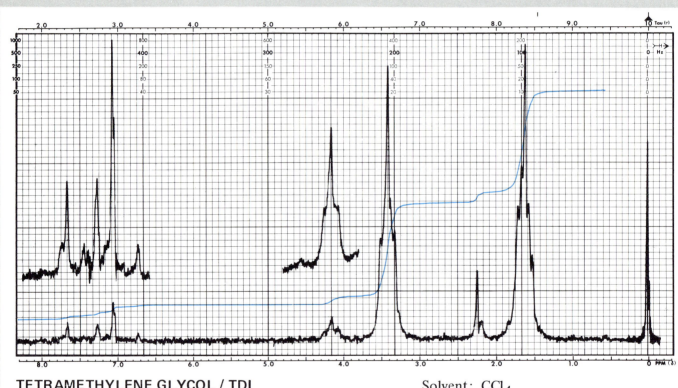

TETRAMETHYLENE GLYCOL / TDI   Solvent: CCl$_4$

COMPLEX STRUCTURE

136

THE EPOXY RESINS

The term epoxide or epoxy resin is applied to a class of relatively low molecular weight compounds which contain one or more terminal glycidyl ether groups ($R-O-CH_2-CH-CH_2$, with an epoxide O bridging the $CH-CH_2$).

Although the number of possible chemical variations is large, most epoxy resins of commercial importance are based on the diglycidyl ether of bisphenol-A (diphenylol propane) or an oligomer of that structure. The synthesis of this important resin is illustrated below:

$$2\,CH_2-CH-CH_2-Cl + HO-\!\!\left\langle\!\!\bigcirc\!\!\right\rangle\!\!\overset{\overset{CH_3}{|}}{\underset{\underset{CH_3}{|}}{C}}\!\!\left\langle\!\!\bigcirc\!\!\right\rangle\!\!-OH \xrightarrow{\ NaOH\ }$$

$$CH_2-CH-CH_2-O-\!\!\left\langle\!\!\bigcirc\!\!\right\rangle\!\!\overset{\overset{CH_3}{|}}{\underset{\underset{CH_3}{|}}{C}}\!\!\left\langle\!\!\bigcirc\!\!\right\rangle\!\!-O-CH_2-CH-CH_2$$

The oligomers, which are usually solids, have the general structure:

$$\left[-O-\!\!\left\langle\!\!\bigcirc\!\!\right\rangle\!\!\overset{\overset{CH_3}{|}}{\underset{\underset{CH_3}{|}}{C}}\!\!\left\langle\!\!\bigcirc\!\!\right\rangle\!\!-O-CH_2-\underset{\underset{OH}{|}}{CH}-CH_2-\right]_n$$

Epoxy resins contain two different types of reaction sites; epoxide rings and pendant hydroxyl groups. The epoxide rings can enter into a variety of ring-opening reactions which lead to chain-lengthened polymers. The reagents employed for this "curing" process include aliphatic diamines, glycols and polyols, aromatic diamines, and polyamides with terminal amine groups. The numerous oxygen and nitrogen functionalities in these polymers provide the epoxides with their remarkable ability to adhere to a variety of surfaces.

The pendant hydroxyl groups, which are formed by the opening of the epoxide rings, can be used to form cross-links or pendant carboxyl groups by reaction with acid anhydrides.

Carboxylic acids are also frequently used to esterify the hydroxyl groups present in the final polymer.

After being "cured" by one of the above reactions, the epoxides are quite solvent resistant and, through opening of the epoxide rings, produce polymers which might be characterized as polyethers of glycerol.

NMR Parameters

Resins which contain terminal glycidyl groups display three characteristic complex multiplets which serve as a "fingerprint" for the identification of these molecules.

1. The epoxide ring methylene group is seen as a multiplet centered at 2.65 ppm.

2. The epoxide ring methine of aliphatic ethers appears as a complex multiplet at 3.15 ppm. In ethers bonded to bisphenol-A, it resonates at 3.3 ppm.

3. The ether methylene group produces a multiplet at 3.6 ppm with aliphatic ethers and at 4.0 ppm for the aromatic (bisphenol-A) ethers.

Since bisphenol-A is a common component in epoxide resins, its bands are often indicative of these polymers. This intermediate is, however, often used in other polymers such as the polysulfones and polycarbonates.

The resonance bands of bisphenol-A in epoxy resins appear as follows:

1. The isopropylidene methyl groups appear as a strong, sharp singlet near 1.6 ppm.

2. The aromatic protons produce a characteristic para substitution pattern (two doublets) which are centered about 7.0 ppm. Due to solvent effects, these para doublets may merge to form an apparent singlet.

Commercial Names

Alfane (66), Araldite (67), Epi-Rez (68), Epotuf (69), Epikote (70), Epon (70), ethoxyline resins.

138

Opening of the epichlorohydrin rings results in the formation of a non-cyclic structure represented by the broad, complex band at 3.65 ppm. Terminal epoxide rings produce the well resolved multiplets in the chemical shift range from 2.5 to 3.3 ppm. These bands are essentially identical to those of the monomeric material which appears on page 221.

Based upon the integration values observed in the spectrum, this sample represents the condensation of four epichlorohydrin units as its average molecular structure.

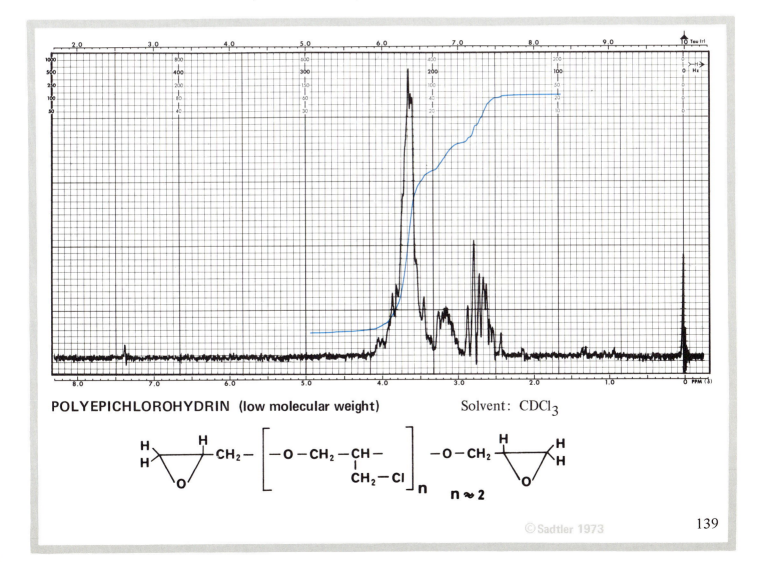

POLYEPICHLOROHYDRIN (low molecular weight)   Solvent: $CDCl_3$

In contrast to the previous spectrum, this sample of polyepichlorohydrin does not display absorbance bands of terminal epoxide rings. The five protons of each polymer unit are accidentally equivalent due to the similar deshielding effect of oxygen and chlorine resulting in a single broad band for all of these hydrogens. Because of its high chlorine content (about 38%) these elastomeric polymers are flame resistant.

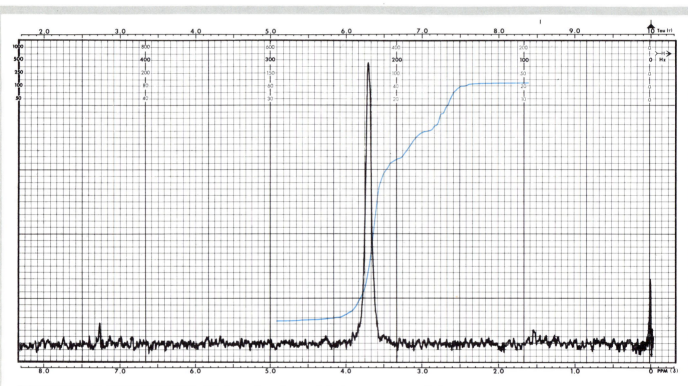

POLYEPICHLOROHYDRIN (high molecular weight)   Solvent: $CDCl_3$

$$\left[-O - CH_2 - CH - \atop \quad\quad CH_2 - Cl \right]_n$$

140

The propylene glycol methylene groups resonate at highest field as a clear doublet at 1.16 ppm. The propylene glycol methylene and methine groups appear as part of the absorbance bands in the chemical shift range from 3.3 to 4.0 ppm. The three ring hydrogens of the glycidyl ether terminating groups are represented by the multiplets in the range from 2.45 to 3.3 ppm. The remaining methylene group resonates at about 3.6 ppm overlapping with the other methylene and methine groups which are adjacent to oxygen atoms. The broad singlet at 3.66 ppm represents the ethylene glycol methylene groups.

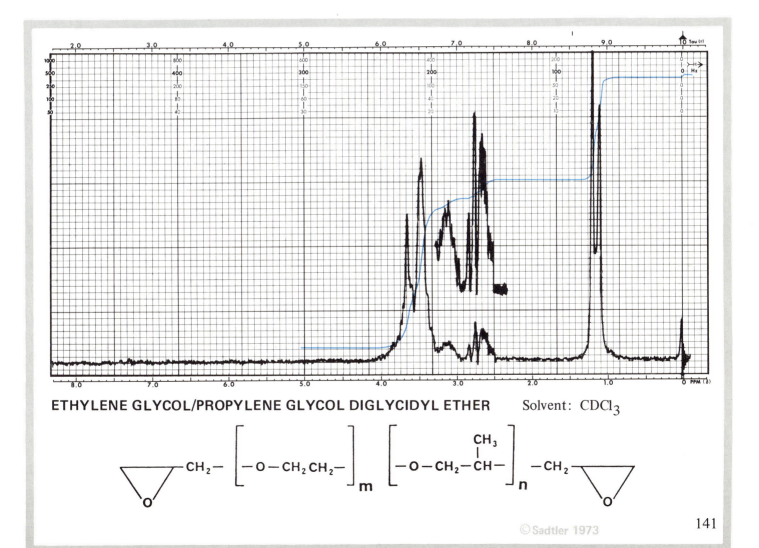

ETHYLENE GLYCOL/PROPYLENE GLYCOL DIGLYCIDYL ETHER   Solvent: $CDCl_3$

©Sadtler 1973

141

EPOXY RESINS

This sample of bisphenol-A diepoxide is the monomer used in the preparation of many of the epoxide resins which appear on the following pages. The gem dimethyl groups produce a characteristic sharp band at 1.61 ppm while the five protons of the epoxide molecule appear as a complex series of bands in the chemical shift range from 2.5 to 4.6 ppm. The aromatic hydrogens produce a clear para substituted pattern consisting of ortho doublets at 6.8 and 7.13 ppm. Opening of the epoxide ring to form polymers results in the loss of absorbance in the range from 2.5 to 3.5 ppm with an attendant increase in the intensity of the absorbance centered at about 4.0 ppm.

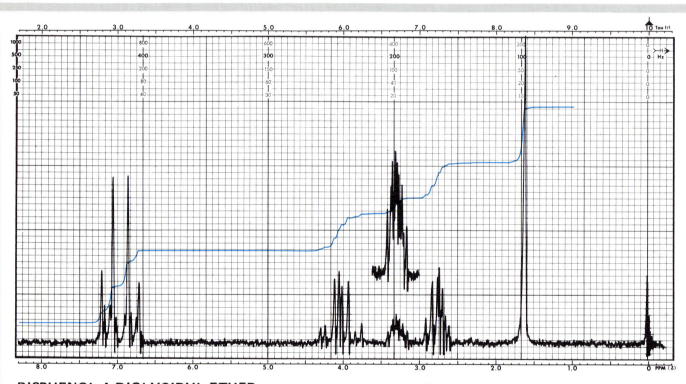

BISPHENOL-A DIGLYCIDYL ETHER

Solvent: CDCl$_3$

142

In the spectrum of this bisphenol-A epoxide resin, polymerization of the monomeric material discussed on the previous page has occurred resulting in the opening of the epoxide ring with a concurrent loss of epoxide absorption in the chemical shift range from 2.5 to 3.5 ppm and an increase in the glycerol resonance centered at about 4.07 ppm. A trace of the glycidyl group is still present as evidenced by the weak band near 2.75 ppm.

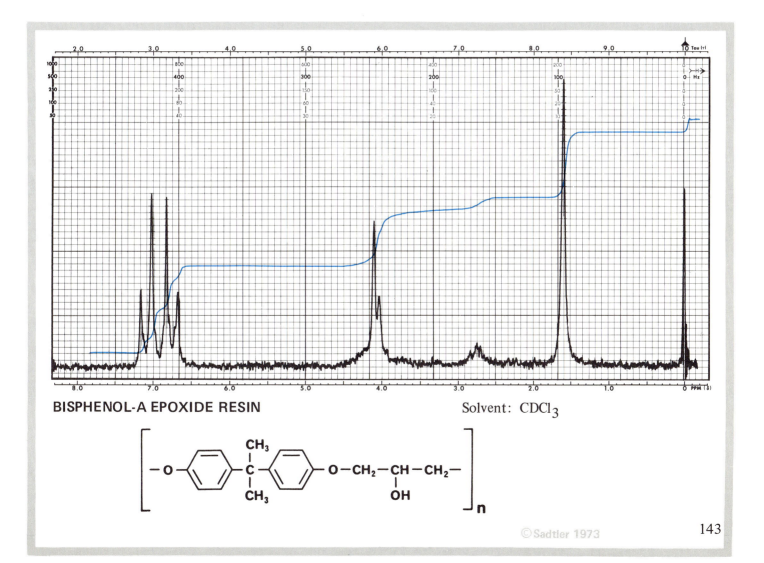

BISPHENOL-A EPOXIDE RESIN

Solvent: $CDCl_3$

143

EPOXY RESINS

The addition of a modifying oil has made this spectrum of an epoxide resin much more complex in appearance than the sample discussed on the previous page. The characteristic bisphenol-A epoxide resonance bands are noted at 1.63, 2.8, 3.3, 4.0 and in the range from 6.6 to 7.3 ppm. The modifying oil is a long chain unsaturated fatty ester, probably a di or triglyceride.

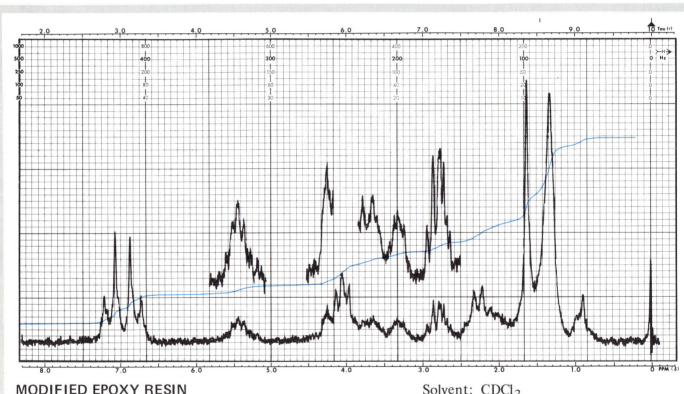

MODIFIED EPOXY RESIN

Solvent: $CDCl_3$

COMPLEX STRUCTURE

This diepoxy resin was prepared by the reaction of two moles of epichlorohydrin with one mole of phenol. In comparison with the previous spectra of bisphenol-A epoxides, this spectrum does not contain the characteristic gem dimethyl band at 1.6 ppm. Additionally, the aromatic proton pattern is that of a mono-substituted rather than para disubstituted benzene ring. Since all five protons of the glycidyl ether terminating group are non-equivalent as well as all five protons of the central glycerol group, the sample produces a very complex series of bands in the chemical shift range from 2.4 to 4.3 ppm.

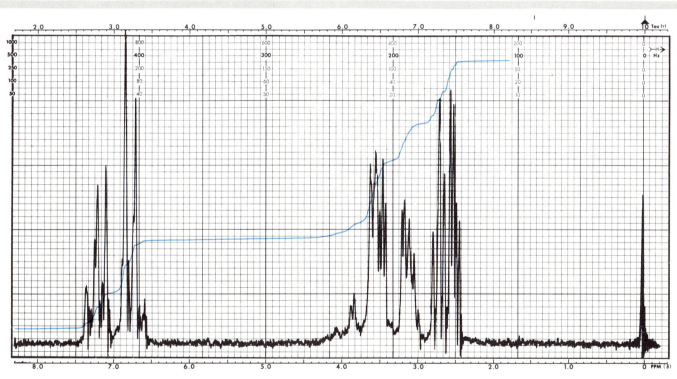

PHENOL DIEPOXIDE RESIN

Solvent: $CDCl_3$

145

The reaction of a phenol epoxide resin such as that scanned on page 145 with a limited amount of form-aldehyde results in the formation of a novolak resin in which the aromatic rings are joined by CH_2 groups. The bridging methylene groups overlap with the epoxide band at about 4.0 ppm thereby increasing its integration value. The additional substituent on the aromatic rings change the aromatic pattern and reduce its integration value. This spectrum depicts such a novolak resin. The aromatic bands at 6.69 and 6.83 ppm represent a para-substitution ortho doublet indicating that the terminal aromatic rings are para substituted.

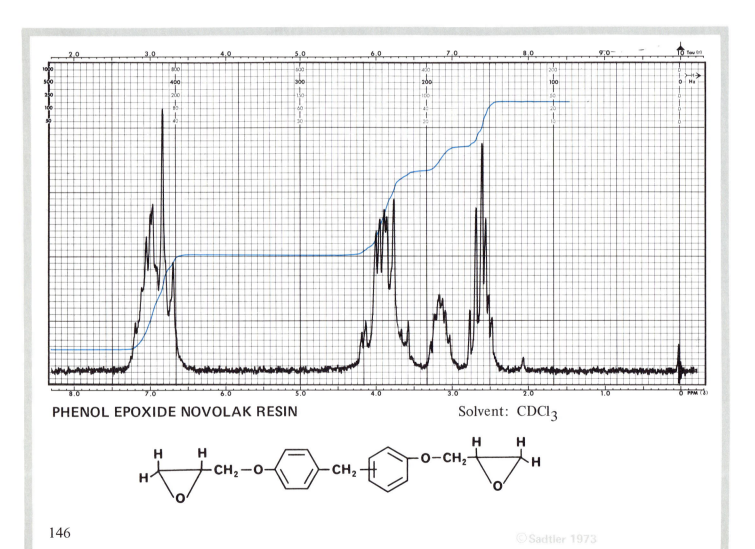

PHENOL EPOXIDE NOVOLAK RESIN   Solvent: $CDCl_3$

THE FORMALDEHYDE RESINS

Phenol-Formaldehyde Resins

The phenol-formaldehyde (PF) resins, as the name implies, are essentially condensation products of phenol and formaldehyde. Alkyl substituted phenols such as cresol, tert-butyl phenol to octyl phenol may also be used but the reactive aldehydic component is almost always formaldehyde. Polymerization can be initiated by either basic or acidic catalysts resulting in two fundamentally different polymer types with important differences in their structural and physical properties.

The base catalyzed polymers were originally named as "Bakelites" (41) although this trademark name is now applied to a wide line of different polymer types. When fully cured, the base-catalyzed PF resins are thermosetting and solvent insensitive. Excess formaldehyde is employed in the polymerization process with the mono-, di- and trimethylol derivatives of phenol being the initial product.

These low molecular weight resins can further polymerize by forming either diphenyl methane linkages or dibenzyl ether bonds.

This reaction is usually terminated when the resins are of intermediate molecular weight and are either viscous liquids or readily soluble solids. They possess unreacted methylol groups and can be cured with heat or the addition of a catalyst yielding cross-linked "network" polymers which are heat insensitive and solvent resistant.

147

The acid-catalyzed resins are prepared with an excess of phenol and the resultant polymer is described as a <u>Novolak</u> resin. Unlike the base-catalyzed materials, these resins have no residual methylol groups but do possess methylene bridges between the phenol rings. The novolak resins require a polyfunctional cross-linking agent for curing which can bond to any unutilized ortho- or para- positions in the resin. Hexa-methylene tetramine is commonly used for this purpose providing the novolak with tri-functional methylene bridges which yield the final cross-linked structure.

NMR Parameters

The NMR spectra of these resins are usually poorly resolved and display broad bands with no indication of any fine structure. The methylene bridging groups between two aromatic rings appear as a broad band at about 3.8 ppm.

The aromatic methylol groups ($\emptyset-CH_2-OH$) and dibenzyl ether methylene groups are similar in chemical shift and resonate as broadened singlets in the chemical shift range from 4.3 - 4.9 ppm.

The aromatic protons appear as a poorly resolved complex multiplet which is usually centered at about 6.65 ppm indicating the strong shielding effect of the phenolic hydroxyl groups on the resonance of the aromatic hydrogens.

Applications

The novolaks are widely used as lacquers and finishes for paper, metal sheet and electrical equipment. These resins are often modified with drying oils.

Other uses include cast and molded resins, laminate boards, adhesives, ion exchange resins and as rubber additives.

Commercial Designations

Deligna (73), Deresit (73), Duron (14), Ervaphene (72), PF resins, phenoplasts, novolaks, novolacs, novolak.

148

Amine-Formaldehyde Resins

The amine-formaldehyde resins may be considered to be chemical cousins of the phenolic resins since in both classes of resins, formaldehyde is used as a reagent which couples the reactive sites of two separate molecules. Additionally, the initial products of both reactions are methylol derivatives which can then react further to form methylene or oxydimethylene bridges ($-CH_2-O-CH_2$) between the other monomer units.

With the amine-formaldehyde resins, the other monomers are either multifunctional amines or amides. The two most commonly used of these are urea and melamine, with compounds such as thiourea, dicyanodiamide, benzoguanamine and aryl sulfonamides being utilized less frequently.

By controlling the ratio of formaldehyde to either urea or leamine, the extent of methylol formation can be controlled so that it is possible, for instance, to produce the polyfunctional hexamethylol melamine in which all of the available hydrogens have been replaced. The resultant monomers can then be polymerized with heat (as well as acids or bases) to form soluble, fusible resins. These resins, finally, can be further polymerized to produce heat stable, insoluble materials.

The molecular structures of such resins and polymers are very complex and are not completely understood. It is known, however, that extensive ring formation occurs leading to highly branched "network" polymers. Although many different ring structures have been proposed, they are all similar in that they contain two or three isolated methylene groups which are substituted in either side by oxygen and/or nitrogen atoms.

NMR Parameters

The amine-formaldehyde resins usually produce poorly resolved NMR spectra. With the exception of modifying groups such as butylated urea, the main feature of their spectra arises from the isolated methylene groups which are formed. Since the type of substitution is variable (e.g. $-O-CH_2-O-$, $-O-CH_2-N-$, $-N-CH_2-N-$) as well as their position in the polymer network, these bands resonate over a very broad region (4.0 - 5.0 ppm) and are poorly resolved. Little information concerning the actual structure of these resins is provided by NMR.

Commercial Designations

Amine-Formaldehyde Resins (general)
Beetle (20), Epok (22), Melan (71), Uformite (34), aminoplasts

Melamine-Formaldehyde Resins
Aminolac (72), Cymel (20), Demilan (73), Diaron (69), Ervamine (72), Melbrite (31), MF resins.

Urea-Formaldehyde Resins
Desurit (73), Kaurit (7), Pollopas (74), UF resins.

Novolak resins which are prepared by the reaction of formaldehyde with an excess of phenol, are linear polymers in which phenolic rings are joined by methylene bridges. Many possible ortho-meta-para isomers are possible resulting in a complex aromatic spectrum and a resin that is usually amorphous. The CH_2 groups isolated between the aromatic rings appear as a broad single peak at about 3.8 ppm overlapping with the phenolic hydrogens which resonate at about 3.60 ppm. Methylol substituents when present appear as bands in the chemical shift range from 4 to 5 ppm.

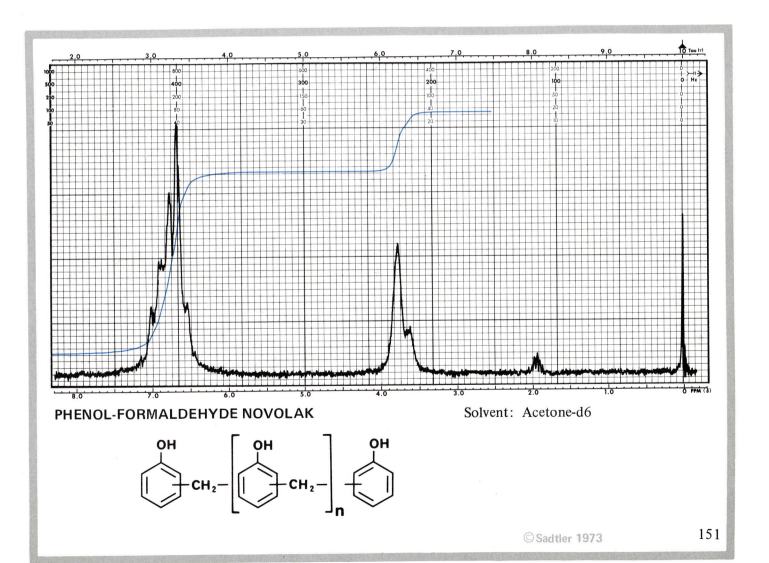

PHENOL-FORMALDEHYDE NOVOLAK　　　　　　Solvent: Acetone-d6

151

FORMALDEHYDE RESINS

The dominant band of this phenol-formaldehyde resin is the intense 9 hydrogen single peak at 1.25 ppm which represents tertiary butyl groups bonded to an aromatic ring. The methylene bridging groups are observed at 3.8 ppm and the aromatic hydrogen band pattern is noted in the chemical shift range from 6.5 to 7.2 ppm. There is no evidence of any significant concentration of methylol (-CH$_2$-OH) substitution of the phenolic rings.

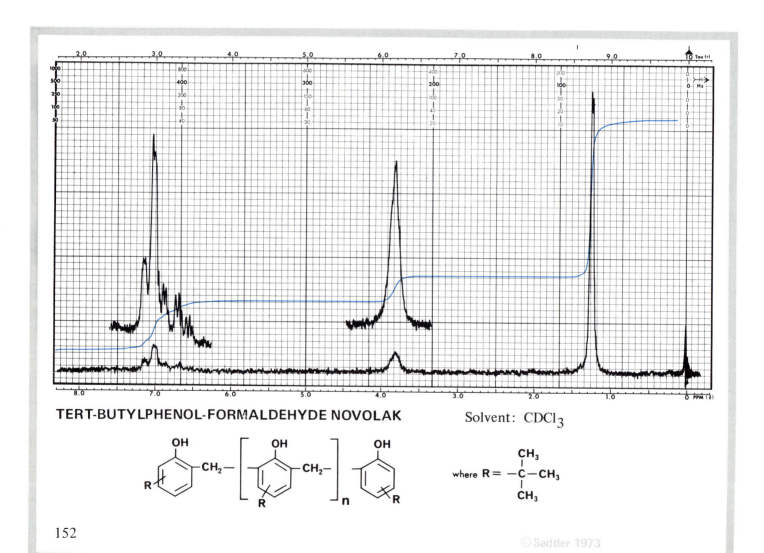

TERT-BUTYLPHENOL-FORMALDEHYDE NOVOLAK   Solvent: CDCl$_3$

Although formed from the same starting materials as the previous sample (tert-butyl phenol and formalde-hyde), this spectrum contains a strong methylol absorbance band at 4.6 ppm. The formation of such methylol (-CH$_2$-OH) groups is favored by the reaction of a phenol with an excess of formaldehyde in the presence of a basic catalyst. These methylol groups form reactive sites in the resin allowing further polymerization or crosslinking of the polymer.

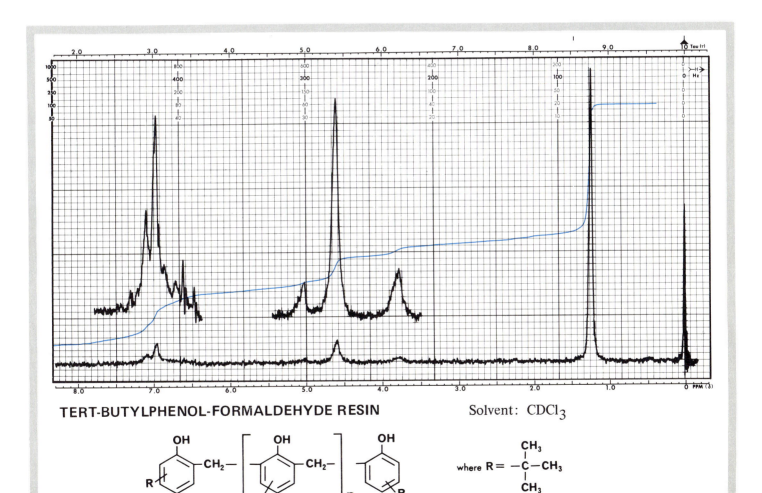

TERT-BUTYLPHENOL-FORMALDEHYDE RESIN   Solvent: CDCl$_3$

where R = $-\overset{CH_3}{\underset{CH_3}{\overset{|}{\underset{|}{C}}}}-CH_3$

153

FORMALDEHYDE RESINS

This phenol-formaldehyde resin displays a unique high field pattern consisting of three sharp singlets with chemical shifts of 0.82 ppm (9 H's), 1.31 ppm (6 H's) and 1.69 ppm (2 H's). These bands arise from the presence of a tetramethylbutyl substituent on the phenolic rings. This starting material is commercially known as "Octyl phenol". The novolak methylene bridging groups resonate at 3.8 ppm as observed in previous spectra. The weak bands at 4.8 and 5.13 ppm probably represent a small percentage of methylol substituents in the sample.

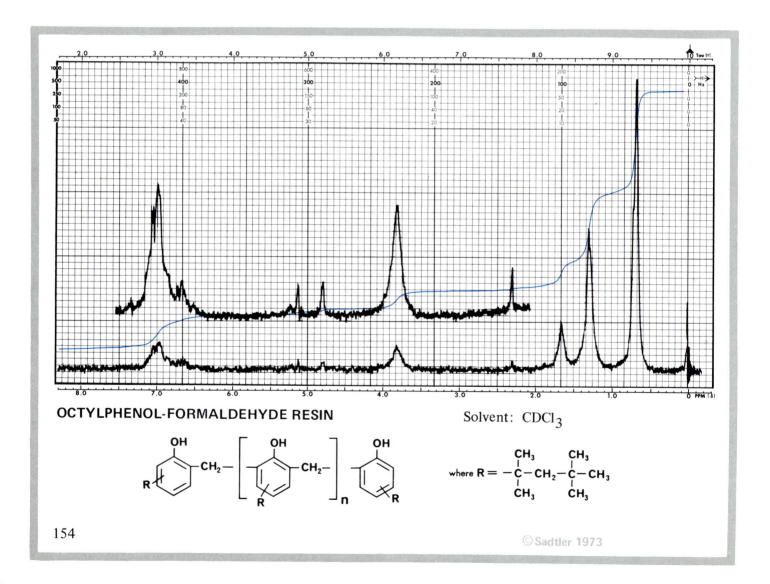

OCTYLPHENOL-FORMALDEHYDE RESIN　　　　　Solvent: $CDCl_3$

154

The butyl ether groups are represented by multiplets at 0.97 ppm (CH_3-), about 1.5 ppm (CH_2-CH_2) and 3.60 ppm (-CH_2-O-). The methylene groups which arise from the addition of formaldehyde to urea are bonded to the urea nitrogen atoms and appear as a broad single peak in the chemical shift range from 4.7 to 5.3 ppm (N-CH_2-O-). The band at 5.7 ppm probably represents the hydroxyl protons of unmodified methylol groups (N-CH_2-OH). The absence of any amide proton absorbance bands indicates that the urea molecules are tetrasubstituted.

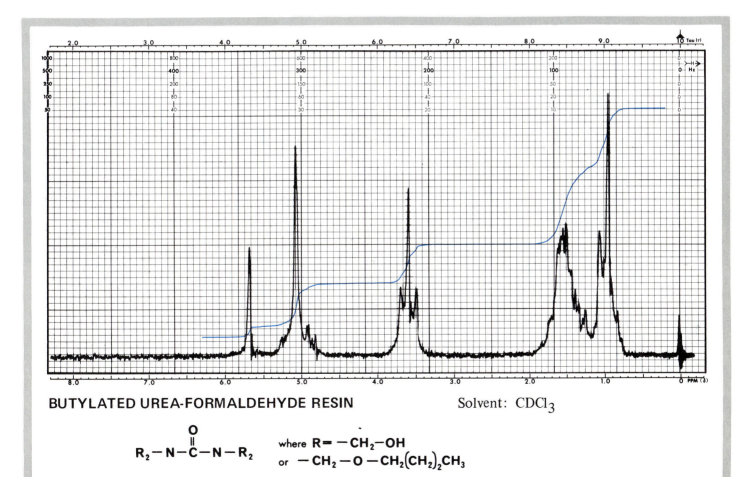

BUTYLATED UREA-FORMALDEHYDE RESIN   Solvent: $CDCl_3$

$$R_2 - N - \overset{\overset{\displaystyle O}{\|}}{C} - N - R_2$$

where R= -CH_2-OH
or -CH_2-O-$CH_2(CH_2)_2CH_3$

155

FORMALDEHYDE RESINS

The para and meta protons of the phenyl substituent resonate as a broad band near 7.4 ppm while the de-shielded ortho protons appear at about 8.35 ppm. The N-substituted methylene ether groups (N-CH$_2$-O-) couple with the adjacent amine protons and appear as a very complex poorly resolved band centered at about 5.27 ppm. The amine protons occur as a very broad band in the chemical shift range from 5.5 to 7.5 ppm. This resin has been modified by the replacement of the methylol -OH groups by butyl ether moieties. The sharp peaks at 2.21, 2.28, and 6.9 to 7.2 ppm indicate the presence of a trace of aromatic solvents in the sample.

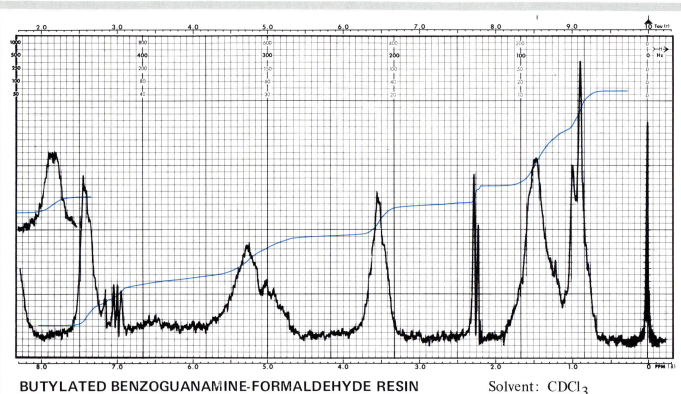

BUTYLATED BENZOGUANAMINE-FORMALDEHYDE RESIN   Solvent: CDCl$_3$

This sample is that of the N-methylol derivative of toluenesulfonamide (CH_3-ϕ-SO_2-NH-CH_2-OH) and unreacted toluenesulfonamide. The observance of three different methyl absorbance bands at high field (2.3 to 2.7 ppm) and the complex aromatic patterns suggest that the starting material may have been a mixture of ortho, meta and para isomers. No evidence is observed in the spectrogram to indicate the presence of 1,3,5-tris(tolylsulfonyl) hexahydrotriazine which is often proposed as a possible component of these resins.

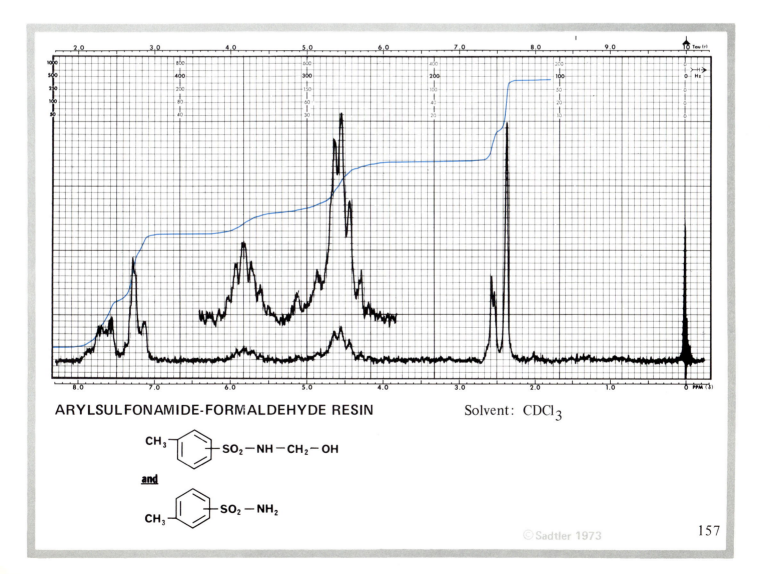

ARYLSULFONAMIDE-FORMALDEHYDE RESIN   Solvent: CCl_3

CH_3—⟨ring⟩—SO_2—NH—CH_2—OH

and

CH_3—⟨ring⟩—SO_2—NH_2

157

THE SILICONE POLYMERS

The silicone polymers are unique among those discussed in this text in that, unlike either the addition or the condensation polymers, the backbones of these compounds ordinarily do not contain carbon atoms. The main chain of the silicones (or siloxanes) consists of alternating silicone and oxygen atoms.

$$R-\underset{\underset{R}{|}}{\overset{\overset{R}{|}}{Si}}-\left[-O-\underset{\underset{R}{|}}{\overset{\overset{R}{|}}{Si}}-\right]_n-R$$

The R-groups may be either alkyl or aryl with the methyl group being the most common substituent. In addition to simple methyl or phenyl substituents, the silanol monomers may be substituted with many other different groups (e.g. fluoroalkyl, ethyl, chlorinated phenyl, alkoxy, acyloxy, chlorine, hydrogen, etc.) which are sometimes utilized as comonomers in the formation of modified silicones.

The vinyl group is an important substituent in certain silicones because it provides a site which can be used for cross-linking of the polymer chains. Another method of achieving cross-linking is to incorporate a certain number of silicon-hydrogen bonds into the initial resin. These silyl hydrogens are susceptible to several cross-linking reactions.

Siloxanes may be formed chemically by the acid or base catalyzed condensation of silanols.

$$2\left[HO-\underset{\underset{R}{|}}{\overset{\overset{R}{|}}{Si}}-OH\right] \longrightarrow HO-\underset{\underset{R}{|}}{\overset{\overset{R}{|}}{Si}}-O-\underset{\underset{R}{|}}{\overset{\overset{R}{|}}{Si}}-OH + H_2O$$

Commercially, these polymers are obtained by the stepwise hydrolysis of organosilyl chlorides to silanols, followed by the polymerization of the silanols to linear or cyclic siloxane oligomers. The silanol monomers may be monofunctional ($R_3-Si-OH$); difunctional ($R_2-Si-(OH)_2$); trifunctional ($R-Si-(OH)_3$) or tetrafunctional ($Si-(OH)_4$). The difunctional monomers are most often used since they lead to simple linear or cyclic polymers. The more highly substituted monomers produce extensively branched compounds which may cause the polymer to gel prematurely and thus cause manufacturing problems. The oligomers are then further polymerized, copolymerized or cross-linked to yield the commercially important silicone fluids, gums and rubbers.

Applications

The commercial silicone rubbers are relatively expensive and are used in those applications in which their servicability at high and low temperatures, resistance to chemical attack, and lack of taste and odor are of special importance. The silicone resins are used as water repellant and non-stick coatings while the fluids are used as lubricants, hydraulic and diffusion pump fluids, and as anti-foam or foam stabilizing agents.

NMR Parameters

Most silicone polymers can be easily recognized by the appearance in their spectra of absorption bands at extremely high field (0.0 - 0.4 ppm) which arise from those methyl groups which are bonded to silicone. The chemical shifts of the alkyl groups in silicone polymers are solvent dependent and additionally vary in chemical shift depending upon their position in the polymer. Few other polymers will contain clearly defined resonance bands at such high a field.

The phenyl groups which are bonded to silicone nuclei in the polymer chain, appear as two broad bands with chemical shifts at about 7.4 ppm (the para and meta hydrogens) and 7.6 ppm (the ortho hydrogens).

An interesting feature which may occasionally be observed, is the reverse appearance of a typical ethyl group when it is bonded to silicone. In this case, the methylene quartet appears at higher field than the methyl triplet. This atypical pattern, in addition to the high field position of these multiplets, is strongly indicative of silicone-carbon bonding.

Nomenclature

The nomenclature of the simple oligomers is usually completely descriptive as indicated by the names and structures shown below.

HEXAPHENYL DISILOXANE

OCTAMETHYL CYCLOTETRASILOXANE

A wide variety of descriptive names are applied to the commercial materials such as organosilicon polymers, polysiloxanes, polydimethyl siloxane, methyl phenyl silicone fluid, silicone (rubber, resin, fluid), alkyl-aryl siloxane polymer.

Commercial Designations

Adrub (75), Eccosil (76), RTV (46), Silaneal (77), Silaprene (43), Silastic (77), Silco-Flex (78), SR-173 (46), Sylgard (77), Viscasil (46).

160

Hydrocarbon groups which are bonded to silicone atoms normally display a strong shift to high field. Methyl groups of polysiloxane polymers normally resonate in the chemical shift range from 0.0 to 0.4 ppm. In the NMR spectrogram of a low molecular weight silicone grease, the methyl groups resonate at 0.09 ppm as a sharp single peak. The absence of any significant absorbance bands in the chemical shift range from 7.0 to 8.0 ppm indicate that this material does not contain any phenyl groups.

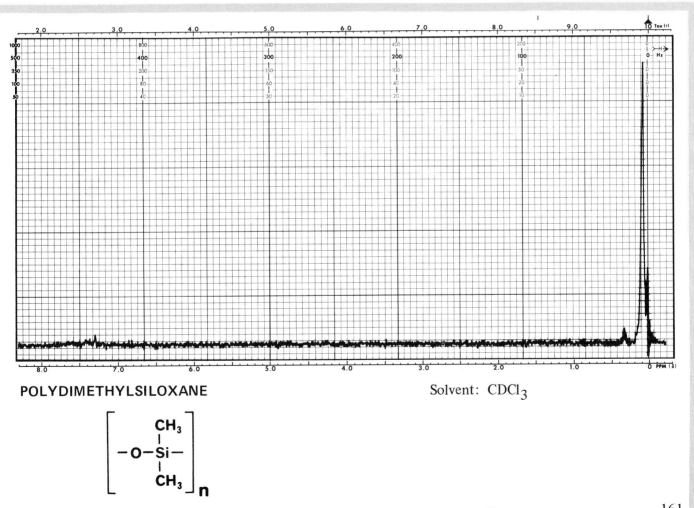

POLYDIMETHYLSILOXANE

Solvent: $CDCl_3$

161

SILICONE POLYMERS

With high molecular weight or crosslinked dimethyl siloxane polymers such as this silicone rubber, the reduced solubility of the polymer requires the use of powerful solvents such as trifluoroacetic acid. The solvent in this case cleaves some of the Si-O-Si linkages resulting in a deshielded methyl band (0.45 ppm) as a decomposition product. The methyl band at highest field (0.19 ppm) is slightly deshielded by this solvent in comparison to the chemical shift observed for a similar polymer examined as a solution in CDCl$_3$ (page 161).

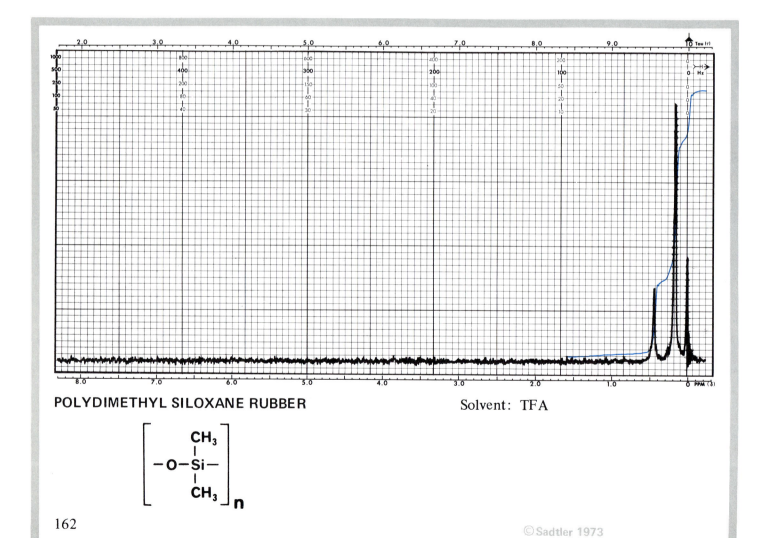

POLYDIMETHYL SILOXANE RUBBER   Solvent: TFA

$$\left[-O-\underset{\underset{CH_3}{|}}{\overset{\overset{CH_3}{|}}{Si}}- \right]_n$$

162

In addition to the ethylene glycol band at 3.68 ppm and the propylene glycol bands at 1.17 and 3.2 to 3.8 ppm, a sharp singlet is observed in this spectrum at very high field (0.1 ppm) which represents the methyl groups bonded to silicon of dimethyl siloxane segments. The minor peaks at 1.3 and 2.07 probably represent impurities or additives. Based upon a comparison of the integration values, the composition of this sample has been estimated to be:

Ethylene glycol	45%
Propylene glycol	43%
Dimethyl siloxane	12%

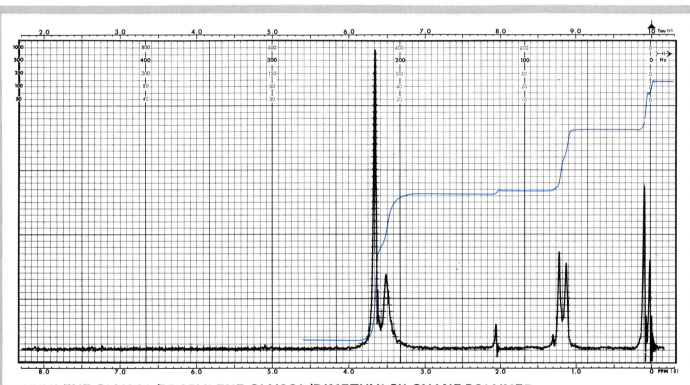

ETHYLENE GLYCOL/PROPYLENE GLYCOL/DIMETHYLSILOXANE POLYMER

Solvent: $CDCl_3$

163

The two sharp singlets at highest field represent methyl groups bonded to silicone in slightly different environments. The multiplet at 0.88 ppm arises from a methylene group bonded to silicone and beta to a trifluoromethyl group ($Si-CH_2-C-CF_3$) and the broadened pentet at 2.02 ppm represents the methylene group beta to silicone and alpha to the trifluoromethyl group. A comparison of the integration values indicates that the ratio of methyl groups to trifluoropropyl groups is 3 to 2.

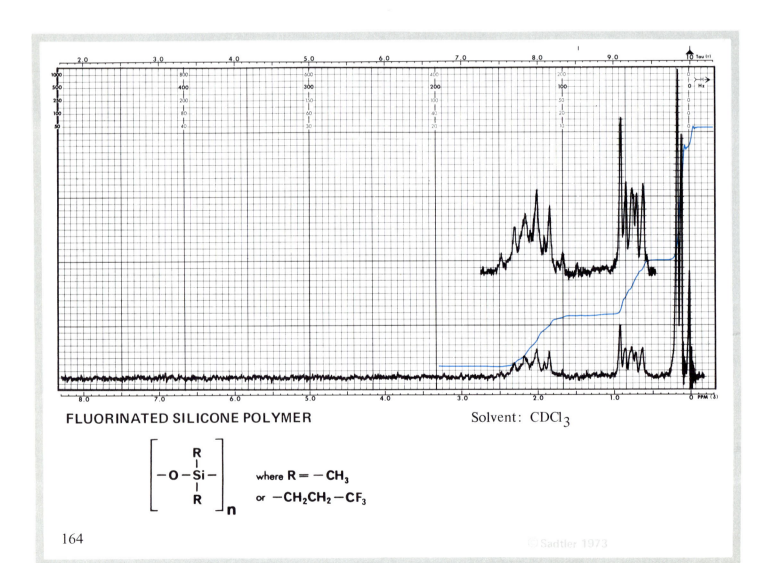

FLUORINATED SILICONE POLYMER   Solvent: $CDCl_3$

$$\left[-O-\underset{\underset{R}{|}}{\overset{\overset{R}{|}}{Si}}- \right]_n$$

where R = $-CH_3$
or $-CH_2CH_2-CF_3$

The presence of other aliphatic groups in addition to methyl groups may produce extremely complex high field patterns as in this spectrum. In addition to methyl siloxane groups (0.08 ppm), the triplet at 0.98 and the quartet at 0.53 ppm indicate the presence of ethyl groups bonded to silicone. The weak singlet which resonates above TMS at (-) 0.07 ppm probably represents ethylene groups bonded to two silicone atoms (Si-CH$_2$-Si). The singlet at 7.21 indicates the presence of a phenyl group bonded to a methylene or methine.

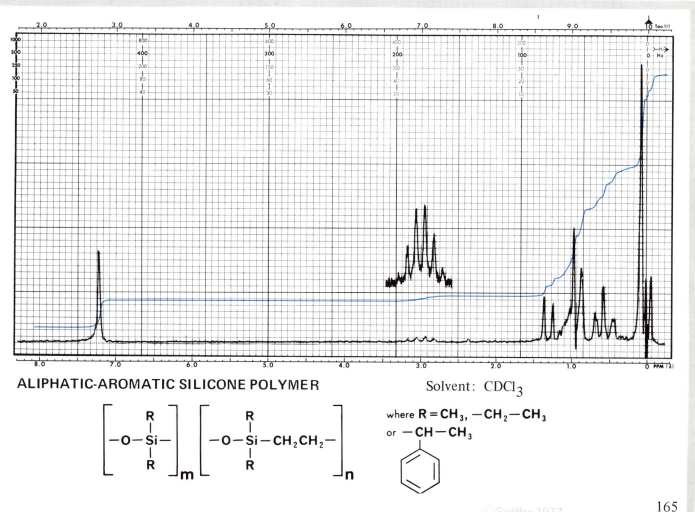

ALIPHATIC-AROMATIC SILICONE POLYMER

Solvent: CDCl$_3$

$$\left[-O-\underset{R}{\overset{R}{Si}}- \right]_m \left[-O-\underset{R}{\overset{R}{Si}}-CH_2CH_2- \right]_n$$

where $R = CH_3$, $-CH_2-CH_3$ or $-CH-CH_3$ (phenyl)

165

SILICONE POLYMERS

The aromatic protons of the chlorinated phenyl groups resonate as a complex series of multiplets in the chemical shift range from 7.1 to 7.8 ppm. The general appearance of the pattern suggests a high percentage of ortho substituted rings. The methyl groups bonded to silicone appear as a complex, broadened band centered at about 0.09 ppm. The presence of the aromatic rings in the polymer result in a preferential deshielding of those methyl groups which are nearby, resulting in several different types of methyl groups with slightly different chemical shifts.

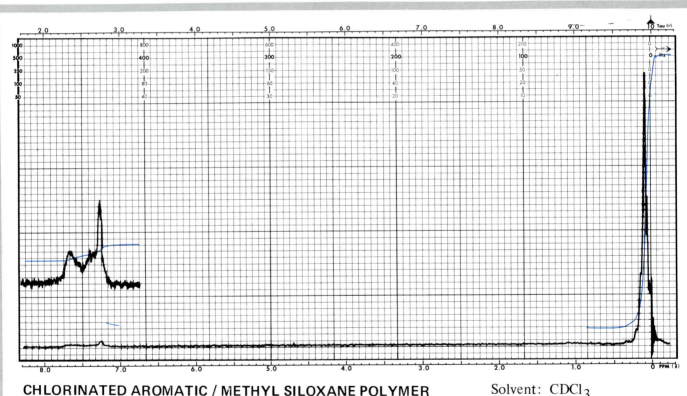

CHLORINATED AROMATIC / METHYL SILOXANE POLYMER   Solvent: $CDCl_3$

COMPLEX STRUCTURE

The aromatic proton pattern and its integration ratio indicates that the aromatic protons represent a mono-substituted benzene ring. The silicone substituent preferentially deshields the ortho protons (7.62 ppm in comparison to the para and meta hydrogens (7.39 ppm). The methyl resonance appears as a complex series of overlapping bands in the chemical shift range from 0.0 to 0.62 ppm. The large number of different chemical shifts observed suggest that the phenyl groups are randomly situated along the polymer chain resulting in many slightly different environments for the nearby methyl groups. The ratio of methyl to phenyl groups is approximately 2 to 1.

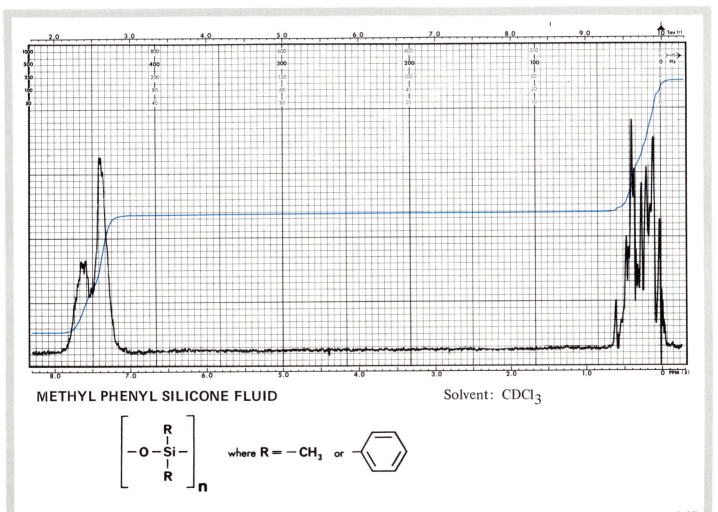

METHYL PHENYL SILICONE FLUID   Solvent: $CDCl_3$

$$\left[-O - \underset{\underset{R}{|}}{\overset{\overset{R}{|}}{Si}} - \right]_n$$

where $R = -CH_3$ or phenyl

In this spectrum of a phenyl methyl silicone polymer, the high field methyl pattern is slightly less complex than that of the previous spectrogram. The methyl groups resonate in the chemical shift range from about 0.0 to 0.3 ppm. The para and meta aromatic protons appear as a complex multiplet centered at 7.33 ppm while the deshielded ortho hydrogens resonate near 7.58 ppm. The ratio of methyl groups to phenyl groups is about 4 to 1.

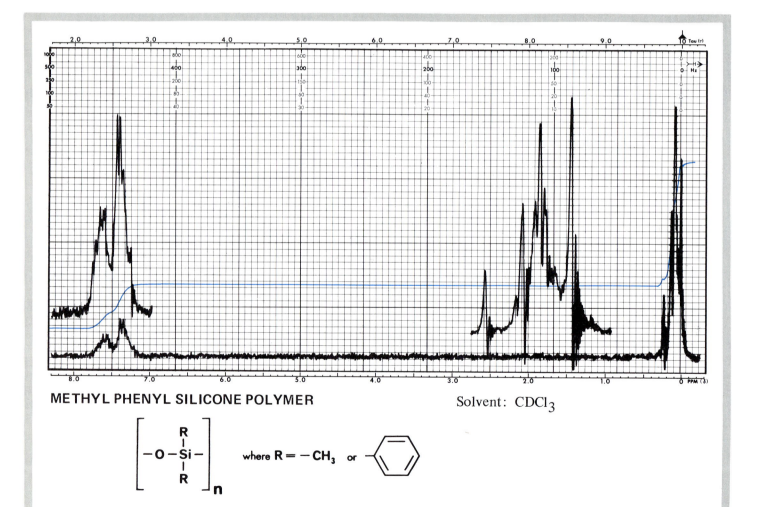

METHYL PHENYL SILICONE POLYMER   Solvent: $CDCl_3$

$$\left[-O-\underset{\underset{R}{|}}{\overset{\overset{R}{|}}{Si}}- \right]_n \quad \text{where } R = -CH_3 \text{ or } \text{—} \langle \bigcirc \rangle$$

168

THE CELLULOSE DERIVATIVES

Cellulose is a naturally occurring polymer which finds wide application in the textile and paper industries. Chemically, it is a polyglucose in which each glucose unit exists in the pyranose (six membered cyclic hemiacetal) form. Each ring is attached to the next through a β-glycosidic linkage.

Many chemical modifications of natural cellulose are possible which greatly increase the use spectrum of this widely available polymer. When unaltered, cellulose is a heat stable, virtually insoluble material which must be processed primarily through mechanical means. When chemically modified however, it may become soluble in water or organic liquids, form films or be cast. In other words, when it is modified, cellulose can become a "plastic".

The reactive sites of the cellulose unit are the two ring hydroxyl groups and the pendant hydroxymethyl group. It is through these three alcohol groups that chemical modification is achieved.

The three main types of chemical modification are:

1. Inorganic esters; nitrate and sulfate
2. Organic esters; acetate, propionate and butyrate
3. Ethers; methyl ethyl, hydroxyethyl, cyanoethyl, carboxymethyl, etc.

The extent of substitution required to achieve pronounced differences in the properties of cellulose is usually quite high (in excess of 70%). In addition to the desired substitution reaction, chemical modification also cleaves some of the glycosidic linkages leading to a significant decrease in molecular weight. As a result, the solubility characteristics of the modified polymer are improved.

The cellulose derivatives are another category of polymers which produce poorly resolved NMR spectra. In addition to the large number of possible isomers, the low solubility of most modified cellulose derivatives and the relatively high viscosity of the resultant sample solutions combine to produce spectra in which the glucose ring hydrogens and pendant methyl groups resonate as a very broad unresolved envelope of absorbance in the chemical shift range from about 2.5 - 6.5 ppm. The ether derivatives resonate over the higher field portion of this range while the organic and inorganic acid esters resonate over the lower field positions.

In theory three types of bands should be visible:

1. The ring methines at lowest field (4 - 6.5 ppm)
2. The pendant methylene groups (3 - 4.5 ppm)
3. The bands which arise from the modifying substituents (0.6 - 4.5 ppm)

Commercial Designations

Cellulose Acetate
Acelon (27), Acetphane (79), Arnel (40), Bexoid (80), Cellon (74), Cellonex (74), acetyl-cellulose, safety celluloid, CA

Cellulose Acetate Butyrate
Cabulite (27), Cobocell (81), CAB

Cellulose Propionate
Celadex (82)

Ethyl Cellulose
Ethocel (3), Ethulon (27), cellulose ethyl ether

Nitrocellulose
Celluloid, collodion, gun-cotton, pyroxylin, cellulose nitrate

The absorbance bands due to the cellulose acetate component of this sample appear as broad, poorly resolved bands at 2.23 ppm and in the chemical shift range from 3.5 to 6.0 ppm. The well resolved multiplets at 1.48, 4.09, 4.53 and 7.74 ppm represent the plasticizer-methylethyl phthalate. The presence of the low field cellulose resonance bands is more clearly indicated by the integration scan than the absorption spectrum.

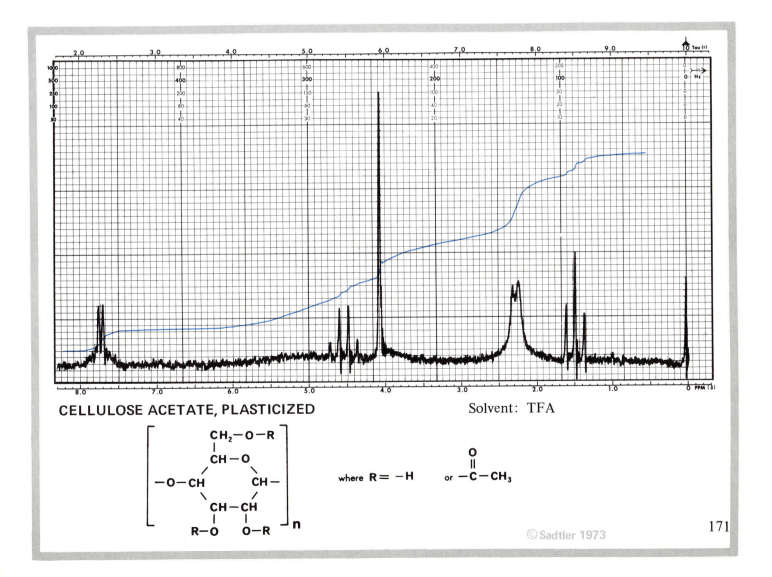

CELLULOSE ACETATE, PLASTICIZED

Solvent: TFA

where R = −H or $-\overset{\overset{\textstyle O}{\|}}{C}-CH_3$

171

CELLULOSE DERIVATIVES

In this spectrogram of cellulose acetate sorbate, the low field cellulose and sorbic acid proton bands are more easily discerned than in the previous spectrogram. The acetate methyl groups resonate at 2.28 ppm. The sharpness of this peak is due in part to the presence of free acetic acid in the sample which overlaps with the cellulose acetate methyl resonance.

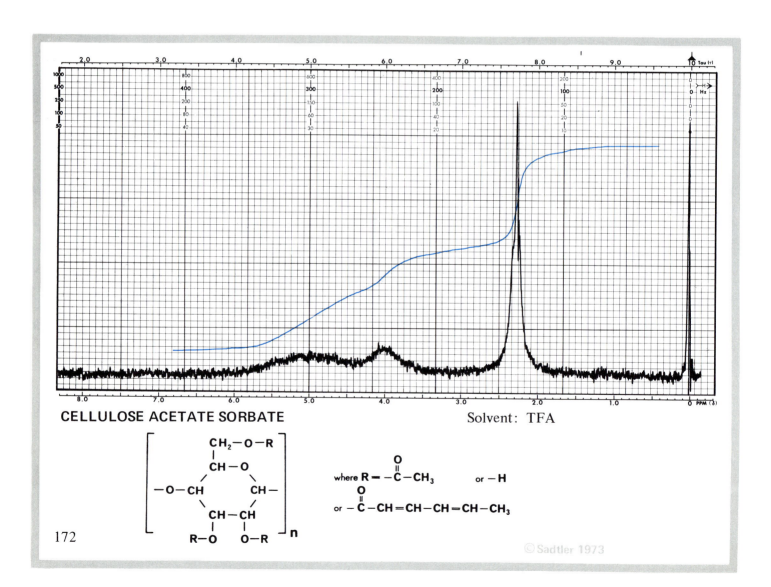

CELLULOSE ACETATE SORBATE

Solvent: TFA

where $R = -\overset{O}{\underset{\|}{C}}-CH_3$ or $-H$

or $-\overset{O}{\underset{\|}{C}}-CH=CH-CH=CH-CH_3$

172

The ethyl methyl group resonates at highest field (1.19 ppm) as a broadened triplet. The adjacent methylene quartet is not clearly observed in the spectrum due to overlap with the cellulose ring hydrogens in the chemical shift range from 2.5 to 5.5 ppm. Based upon a comparison of the integration values, approximately 74% of the cellulose hydroxyl protons have been replaced by ethyl groups.

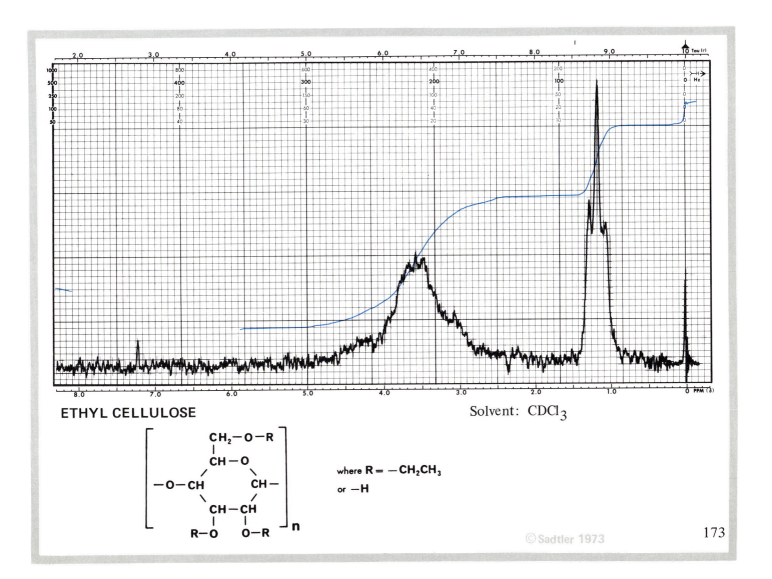

ETHYL CELLULOSE

Solvent: $CDCl_3$

where $R = -CH_2CH_3$

or $-H$

173

CELLULOSE DERIVATIVES

The propionate methyl groups appear as two overlapping triplets centered at about 1.13 ppm. The adjacent methylene groups resonate at 2.28 ppm as a very broadened multiplet which does not display the expected quartet multiplicity. As in previous spectra, the cellulose ring hydrogens resonate at lowest field as a very broad unresolved envelope of absorbance in the chemical shift range from 2.5 to 5.5 ppm.

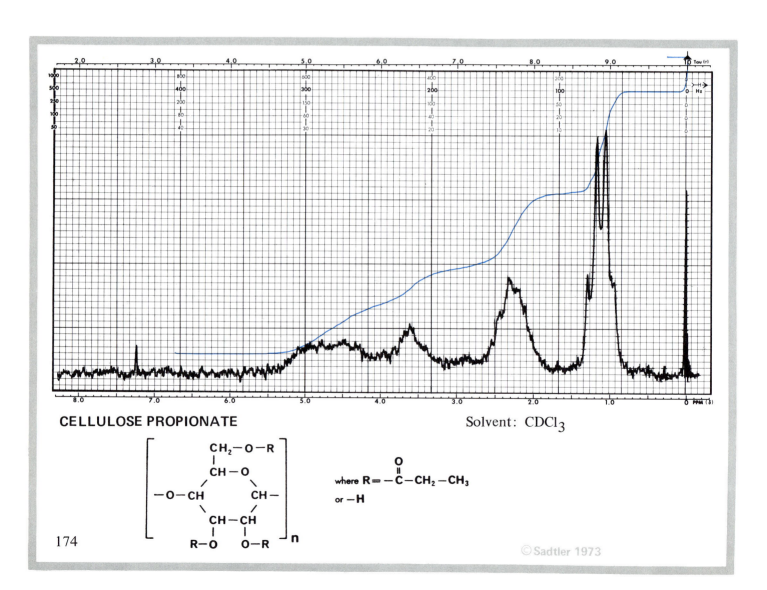

CELLULOSE PROPIONATE

Solvent: $CDCl_3$

where $R = -\overset{\displaystyle O}{\underset{\displaystyle \|}{C}}-CH_2-CH_3$

or $-H$

174

The butyric acid moieties are represented by a complex series of multiplets in the chemical shift range from 0.8 to 2.8 ppm. The methylene group bonded to the ester carbonyl resonates at about 2.13 ppm, the center methylene group appears at 1.67 ppm and the terminal methyl groups are observed as several overlapping triplets at 0.98 ppm. The cellulose ring hydrogens appear as a broad, poorly resolved band of absorbance in the chemical shift range from 3.0 to 6.0 ppm.

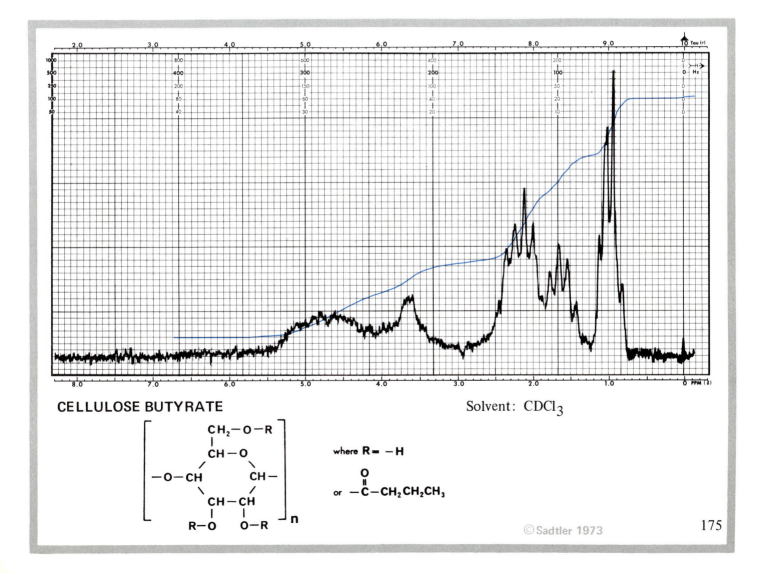

CELLULOSE BUTYRATE

Solvent: $CDCl_3$

where R = —H

or $-\overset{\overset{\text{O}}{\|}}{C}-CH_2CH_2CH_3$

175

All of the protons of nitrocellulose are deshielded and appear as a broad low absorbance band in the chemical shift range from 2.7 to 6.2 ppm. The weak doublet at high field represents the methyl groups of isopropanol which is present in the sample as a wetting agent to reduce the risk of fire or explosion. The band at 2.54 ppm is the solvent impurity DMSO-d5.

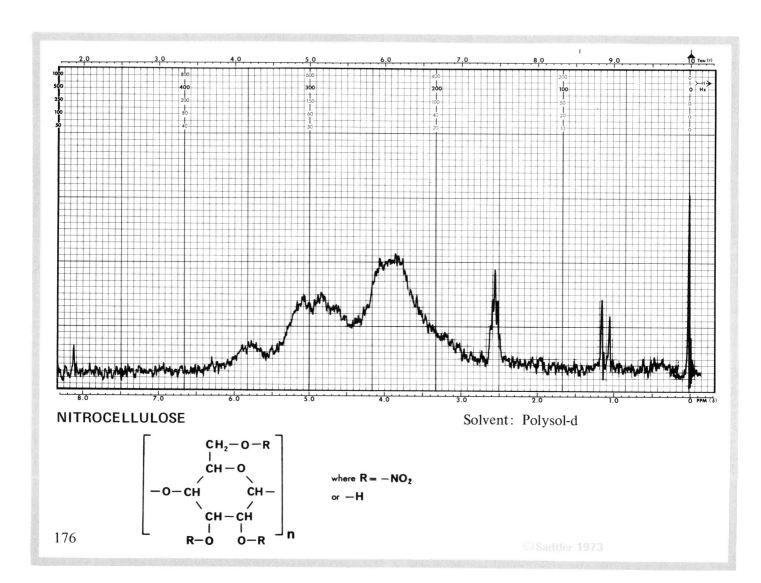

NITROCELLULOSE

Solvent: Polysol-d

where R = $-NO_2$

or $-H$

176

©Sadtler 1973

MISCELLANEOUS POLYMERS

The polymers in this section are generally of indeterminate composition, which produce poorly resolved spectra that do not readily yield to simple interpretation and proton assignments. Many of these polymers are derived from natural sources such as castor oil and pine oil extracts which are themselves complex mixtures of several components. In general, these polymers produce spectra which are characteristic of the polymer type but provide little real information concerning their structural composition.

With only a few exceptions, the polymers in this group are formulated from five basic materials:

a. Triglyceride unsaturated ester polymers from castor oil (glycerol triricinoleate)

$$\left[CH_3-(CH_2)_5-\underset{OH}{CH}-CH_2-\overset{H}{C}=\overset{H}{C}-CH_2-(CH_2)_5-CH_2-\overset{O}{C}- \right]_3 \left[-O-CH_2-\underset{O-}{CH}-CH_2-O- \right]$$

b. Rosin ester gums containing a high abietic acid content

c. Terpene resins from α- and β- pinene

177

d. Cyclohexanone resins

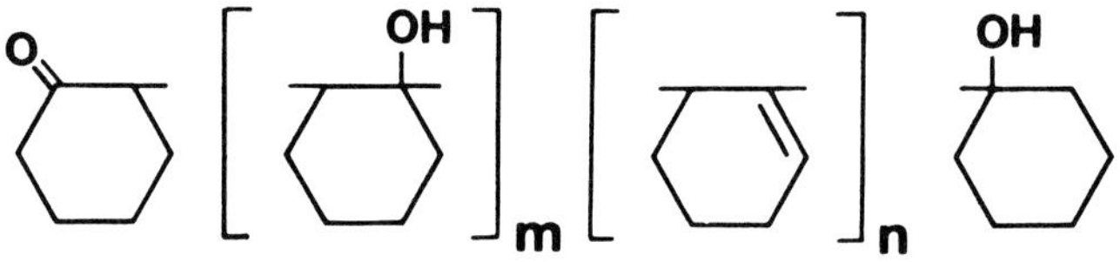

e. Coumarone and/or Indene resins

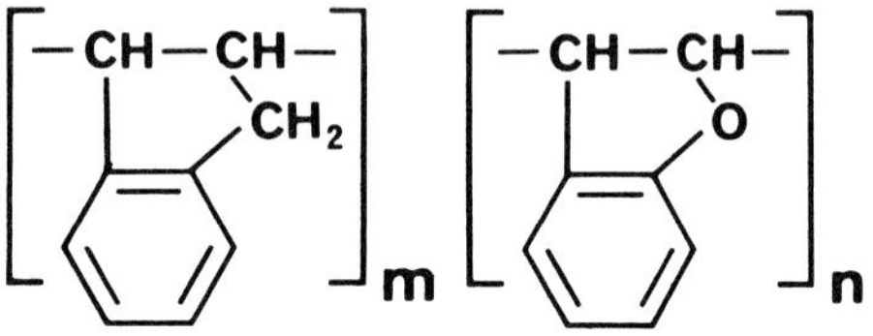

As the structures drawn above indicate, polymers derived from these starting materials produce complex, poorly-resolved NMR spectrograms. In addition, these materials are often extensively modified with other organic compounds, such as phenol, in order to alter the characteristics of the final product.

178

The major absorbance band in this spectrum, the intense singlet at 1.25 ppm, represents the resonance of the sixteen central methylene groups in the octadecyl chain. The terminal methyl protons appear at 0.89 ppm as a distorted triplet. The CH_2 groups bonded to the nitrogen atoms are represented by the broadened triplet near 3.1 ppm while the chain methine groups resonate at 4.85 ppm overlapping with the carbamate NH protons. The broad band near 1.99 ppm is probably due to the presence of residual acetate groups from the starting material, polyvinyl alcohol.

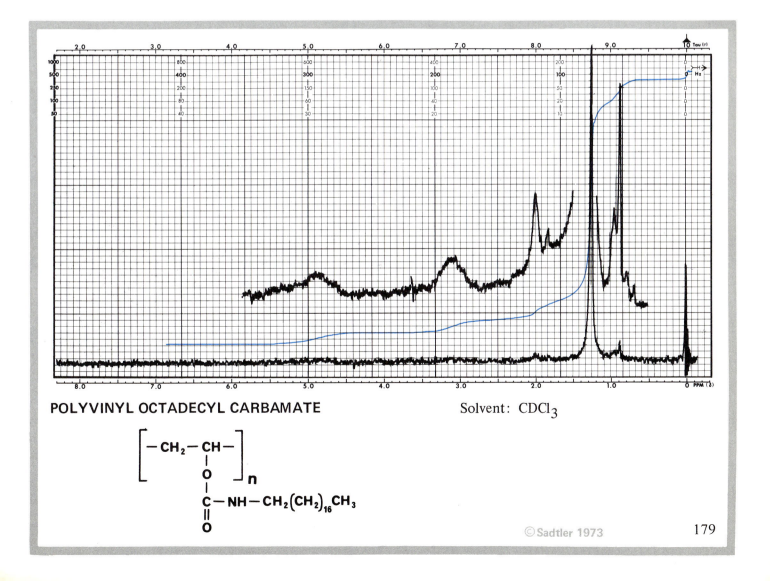

POLYVINYL OCTADECYL CARBAMATE

Solvent: $CDCl_3$

$$\left[-CH_2-CH- \right]_n$$
$$O$$
$$|$$
$$C-NH-CH_2(CH_2)_{16}CH_3$$
$$\|$$
$$O$$

179

The vinyl toluene used to modify this "drying oil" is probably the 3 to 1(meta-para)methyl styrene commercial product. The aromatic protons produce two bands at lowest field near 6.85 and 6.38 ppm. The toluene methyl groups resonate at 2.2 ppm as a broad single peak. The bands at 2.84 ppm, 4.14 ppm and 5.32 ppm indicate that the vegetable oil present in the sample is the triglyceride of a polyunsaturated fatty acid such as linoleic acid.

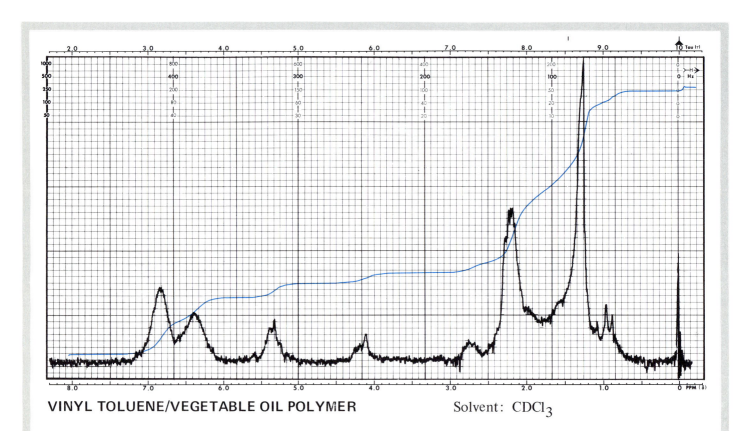

VINYL TOLUENE/VEGETABLE OIL POLYMER   Solvent: CDCl$_3$

COMPLEX STRUCTURE

With the exception of the bands at 2.71 and 4.8 ppm, all of the absorption bands of this spectrum are attributable to the major component of castor oil, glycerol triricinoleate. The single peak at 2.71 ppm suggests that the "polymerization" of this oil was accomplished by the addition of epoxide groups to the castor oil starting material. Epoxidized and chemically polymerized oils such as this are used as plasticizers in adhesives and coating materials.

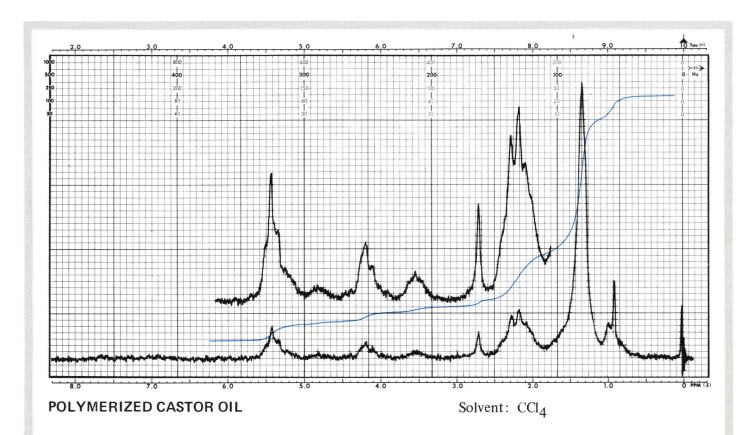

POLYMERIZED CASTOR OIL   Solvent: CCl_4

COMPLEX STRUCTURE

Although similar in general appearance to the previous spectrum, the absence of the band at 3.6 ppm in this spectrum indicates that a reaction at the ricinoleic acid hydroxyl group has occurred. The methine proton adjacent to the OH group has apparently been shifted to lower field (5.0 to 6.5 ppm), suggesting formation of an ester functional group at this site. The characteristic "epoxide" band is noted at 2.72 ppm.

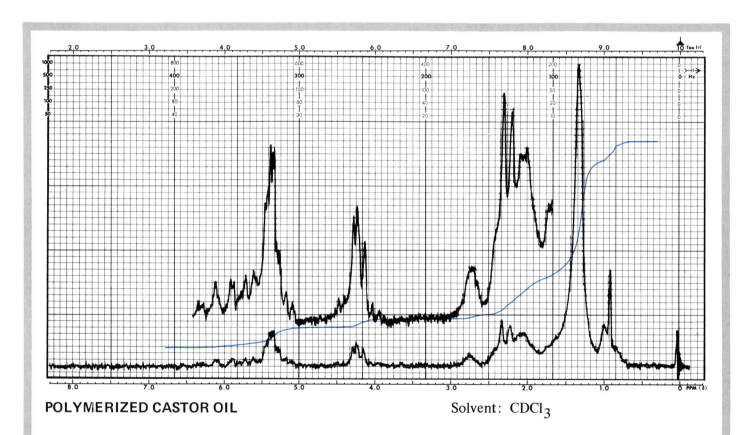

POLYMERIZED CASTOR OIL　　　　　　　　　　　Solvent: CDCl$_3$

COMPLEX STRUCTURE

182

The bands at 0.90, 1.30, 2.2, 4.2 and 5.4 ppm arise from the major component of castor oil, glycerol tri-ricinoleate. The polymerization of this oil has been accomplished by reaction of the fatty acid hydroxyl groups with a diisocyanate which creates a urethane linkage. The bands at 3.6, 4.8 and from 6.5 to 7.8 ppm result from the addition of the aromatic diisocyanate TDI, a mixture of 2,4 and 2,6-tolylene diisocyanate. This polymerization converts castor oil into a drying oil which upon exposure to air will further polymerize to form an elastic film.

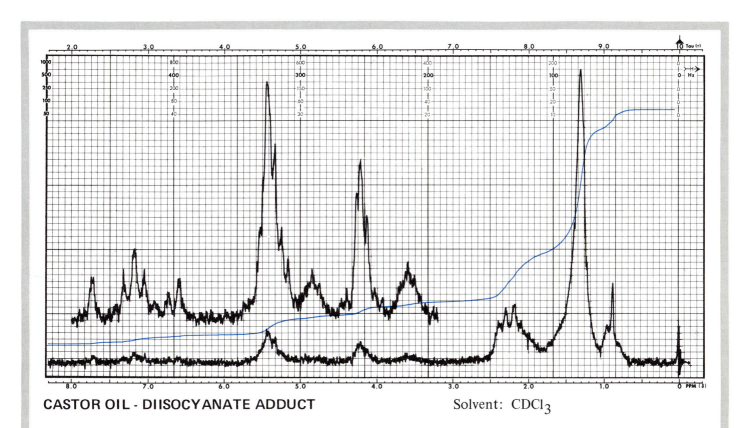

CASTOR OIL - DIISOCYANATE ADDUCT   Solvent: $CDCl_3$

COMPLEX STRUCTURE

The complex series of overlapping multiplets which appear in the chemical shift range from 0.6 to 2.6 ppm arise from the aliphatic protons of rosin acid of which abeitic acid is a major component. The deshielded proton bands which arise from the rosin acid moiety appear at 2.81, 5.30, 5.74, 6.90 and 7.50 ppm. The alcohol used in the preparation of this ester resin is not readily characterizable but is represented by the bands in the chemical shift range from 3.8 to 5.2 ppm. These brittle, readily soluble resins are used in the formulation of varnishes and lacquers.

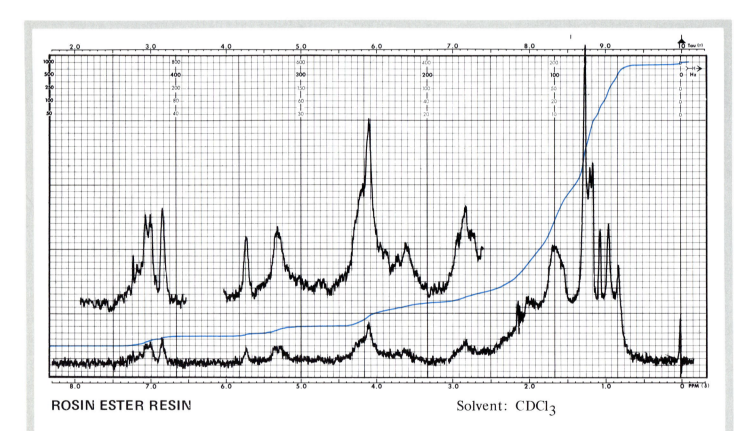

ROSIN ESTER RESIN　　　　　　　　　　　　　Solvent: $CDCl_3$

COMPLEX STRUCTURE

Although similar in general appearance to the previous spectrum, several differences in the peak heights of the high field bands suggest that the rosin acid utilized in the formation of this ester resin possesses a slightly different composition than that of the sample scanned on page 184. The alcohol used to form this ester is pentaerythritol which contains four equivalent methylene groups which resonate as a single broad peak near 4.08 ppm.

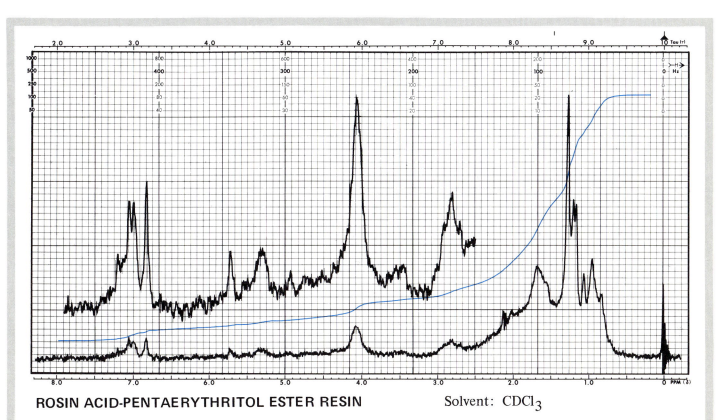

ROSIN ACID-PENTAERYTHRITOL ESTER RESIN   Solvent: $CDCl_3$

COMPLEX STRUCTURE

185

The strongest band in the spectrum of this beta-pinene resin is the broad peak at 0.95 ppm which represents the geminal dimethyl groups. Some hydrogenation of the sample has occurred since the integration value of the olefinic hydrogen band at 5.42 indicates that many of the rings contain no unsaturation. The remaining methylenes and methines of the general structure drawn below resonate over the chemical shift range from 0.6 to 3.5 ppm.

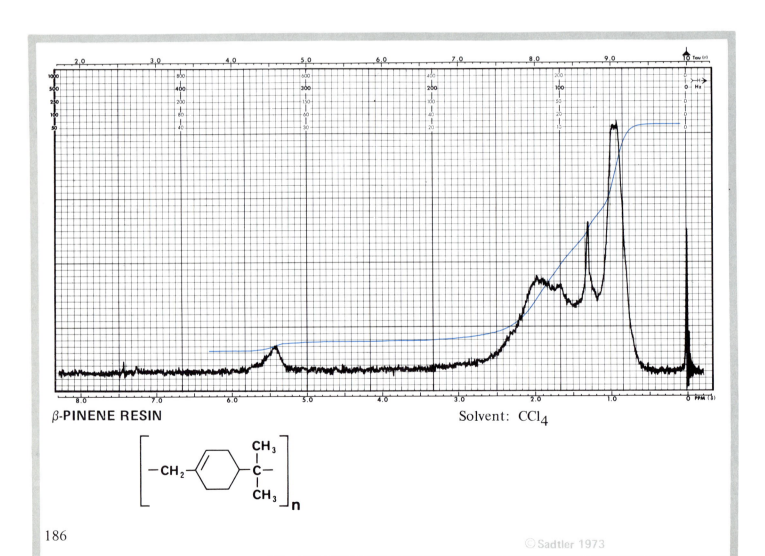

β-PINENE RESIN

Solvent: CCl$_4$

The broadened doublet at 1.68 ppm characterizes this polyterpene as a polymer of alpha-pinene rather than beta-pinene. A similar band is observed in the reference spectrogram of dipentene (limonene) which is scanned on page 245. The first stage in the polymerization of alpha-pinene involves the conversion of the monomer into dipentene followed by polymerization of this intermediate. The broad band at 5.33 ppm represents the resonance of the olefinic hydrogens in the cyclohexene ring system.

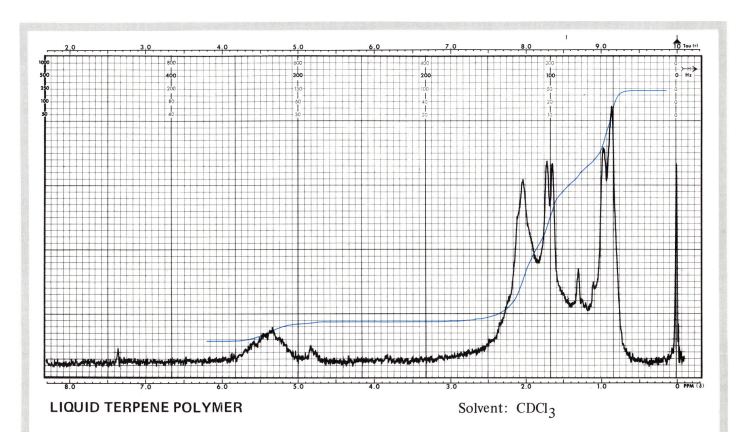

LIQUID TERPENE POLYMER   Solvent: $CDCl_3$

COMPLEX STRUCTURE

187

The appearance of a broad single band near 0.95 ppm and the absence of a broadened doublet at 1.68 ppm indicate that this polyterpene resin is comprised predominantly of beta-pinene units and contains little or no polymerized alpha-pinene groups. The olefinic hydrogens present in the polymer resonate at 5.33 ppm as in previous spectra. The band at about 6.9 ppm suggests that there may be a trace of rosin acid in this sample (see page 247).

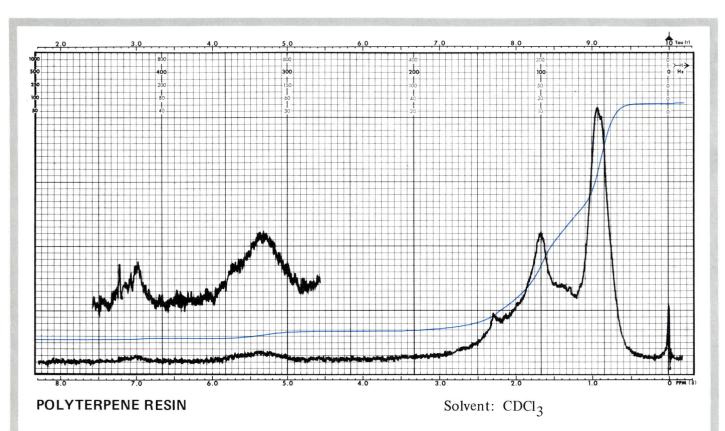

POLYTERPENE RESIN   Solvent: $CDCl_3$

COMPLEX STRUCTURE

The complex series of bands which resonate at high field in the chemical shift range from 0.5 to 3.1 ppm are characteristic of many polyterpene resins. The peak at 0.98 ppm appears in the NMR spectra of most pinene resins. As with the previous phenol-modified resins and polymers, the phenolic aromatic protons resonate as a complex pattern in the range from 6.2 to 7.2 ppm due to the strong deshielding effect of the oxygen substituent. The absence of a significant absorbance band near 5.3 ppm indicates that this sample contains very little residual olefinic unsaturation.

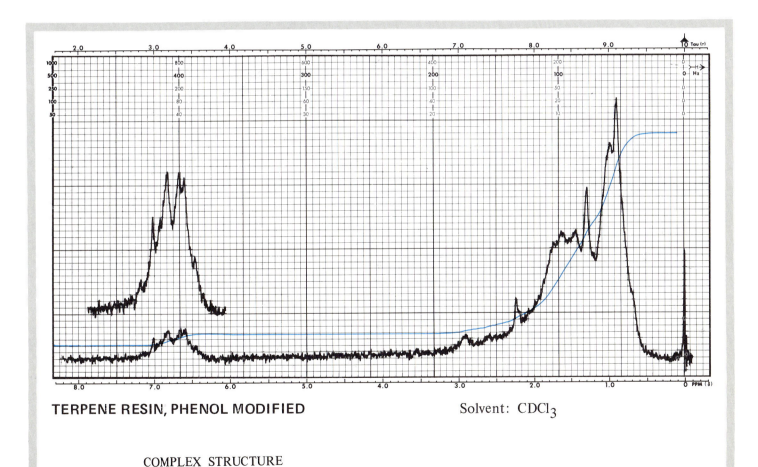

TERPENE RESIN, PHENOL MODIFIED   Solvent: $CDCl_3$

COMPLEX STRUCTURE

189

The high field absorption pattern of this phenol-modified terpene resin differs from those of the previous spectra in that there is no strong band at 0.9 ppm. The absence of such a band indicates that the starting material contained only low percentages of alpha- and beta-pinene. The presence of a sharp peak at 1.28 ppm is indicative of fatty acid methylene groups and is similar in this respect to the reference spectra of tall oil (page 246) and rosin acid (page 247). The phenolic aromatic protons are seen in the chemical shift range from 6.5 to 7.5 ppm.

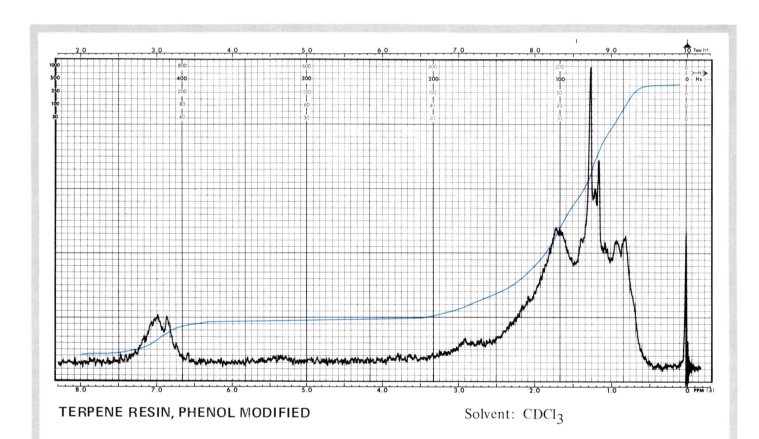

TERPENE RESIN, PHENOL MODIFIED   Solvent: CDCl$_3$

COMPLEX STRUCTURE

The coumarone-indene resins produce a very poorly resolved spectrogram typical of polymers containing alicyclic ring structures. The aromatic protons of both structures possess similar chemical shifts and overlap to form a broad band centered at about 7.2 ppm. The very weak band at about 4.5 ppm, which represents the coumarone methine adjacent to the oxygen atom, indicates that the polycoumarone content of this sample is quite low. The other methylenes and methines of both structures overlap in the chemical shift range from 0.5 to 4.0 ppm as a broad low envelope of absorbance.

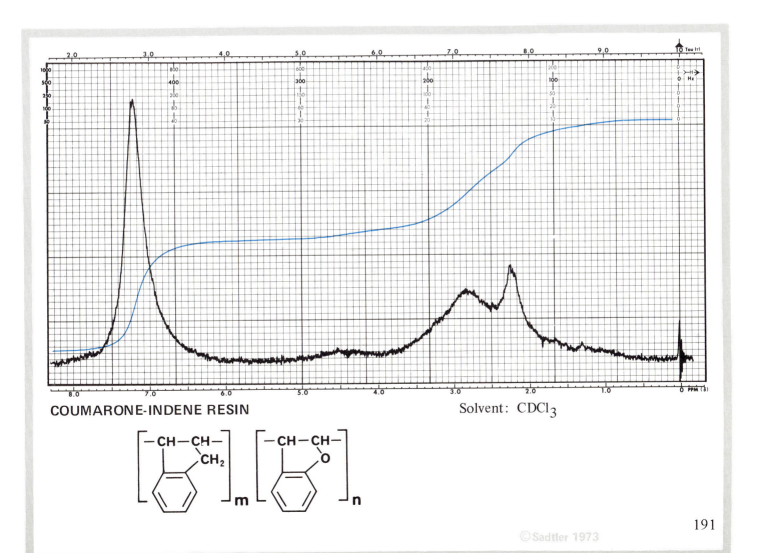

COUMARONE-INDENE RESIN   Solvent: $CDCl_3$

191

MISCELLANEOUS RESINS

As the molecular representation indicates, cyclohexanone condensation resins contain a number of different ring systems resulting in a very poorly resolved spectrogram. The major band is centered at about 1.6 ppm and represents the ring methylene groups which are not adjacent to keto carbons nor double bonds. The methylenes which are adjacent to these groups appear as several weak bands in the chemical shift range from 2.2 to 3.7 ppm. The very weak band at 5.6 ppm probably represents the olefinic protons in the ring. Protons of this type appear to be present at a very low concentration.

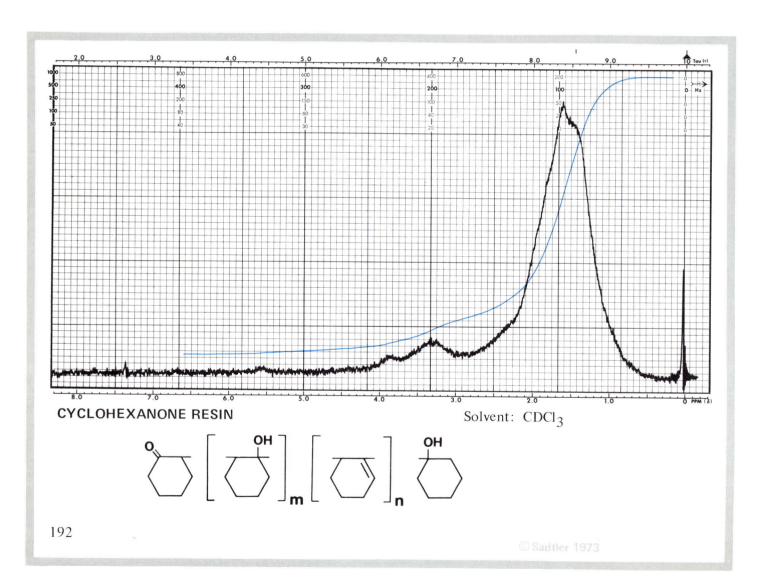

CYCLOHEXANONE RESIN

Solvent: CDCl$_3$

Modification of this cyclohexanone resin by the addition of phenol to the molecular structure results in the appearance of a complex series of multiplets in the chemical shift range from 6.4 to 7.3 ppm which represent the phenol aromatic hydrogens. The band at highest field is essentially unchanged in shape or position (1.65 ppm). The absorption band at 3.8 ppm probably represents the hydrogen remaining at the site of phenol addition.

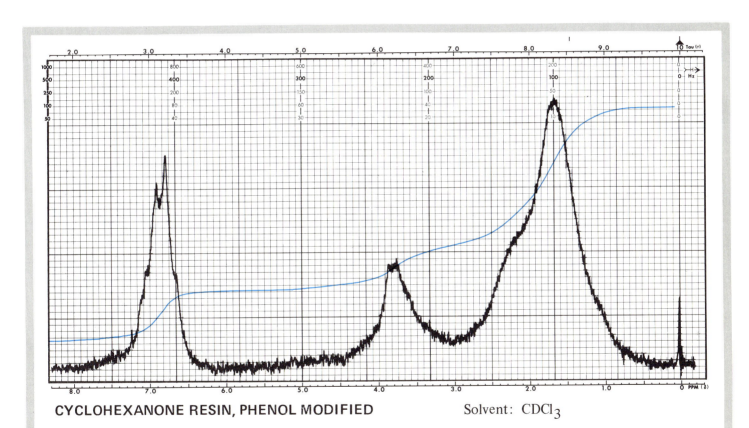

CYCLOHEXANONE RESIN, PHENOL MODIFIED   Solvent: $CDCl_3$

COMPLEX STRUCTURE

MONOMERS

The monomers and reference compounds whose NMR spectra appear on the following pages have been included as examples which illustrate the appearance of the resonance bands of the starting materials, cross-linking agents and additives which may be observed in the NMR spectra of polymers and resins.

These spectra are grouped by compound type and are listed in approximately the same order as the corresponding polymers which are presented on the previous pages of this volume, i.e.,

Monomeric isoprene produces a distinctive set of eight absorbance bands. The methyl group appears at highest field as a three proton broadened singlet at about 1.85 ppm. The two hydrogens which are cis and trans to the methyl group are accidentally equivalent and appear as a single broad band at 4.99 ppm overlapping with the two terminal vinyl protons from the other end of the molecule. The cis/trans doublet of doublets at lowest field represents the olefinic proton in the center of the molecule.

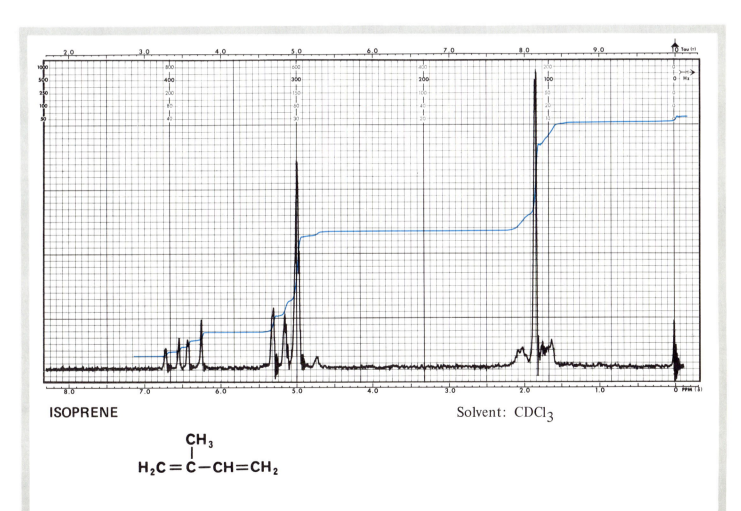

ISOPRENE

Solvent: $CDCl_3$

$$H_2C = \underset{\underset{CH_3}{|}}{C} - CH = CH_2$$

195

In the NMR spectrum of styrene, the five aromatic protons resonate at lowest field as a complex multiplet near 7.15 ppm. At slightly higher field, the hydrogen on the carbon which is alpha to the phenyl group resonates at 6.6 ppm while the two hydrogens on the beta carbons absorb at 5.1 and 5.6 ppm. Because styrene is often used as a crosslinking agent, monomeric styrene may be observed as an additive in the NMR spectra of some low molecular weight polymers.

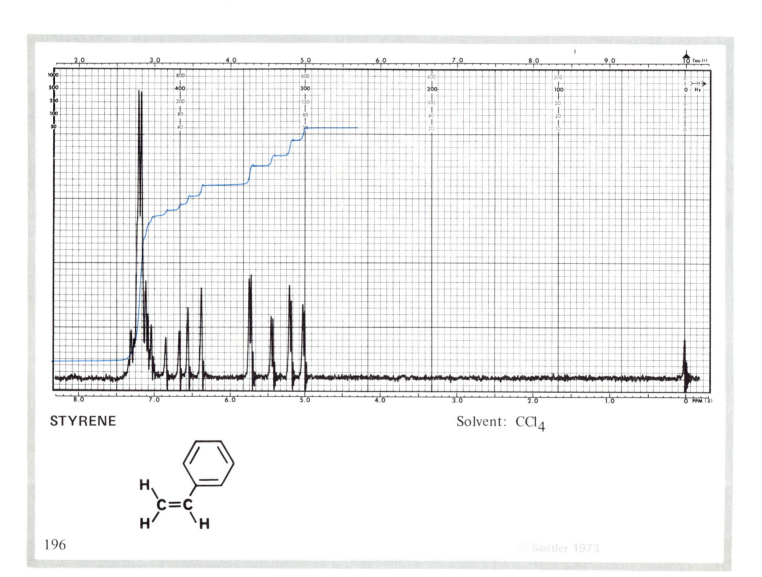

STYRENE

Solvent: CCl_4

The acrylonitrile monomer contains only three vinyl protons which form a complex pattern in the chemical shift range from 5.4 to 6.5 ppm which is similar to those produced by the acrylate monomers. Upon polymerization, the olefinic hydrogens become aliphatic in character and as a result are shifted to higher field.

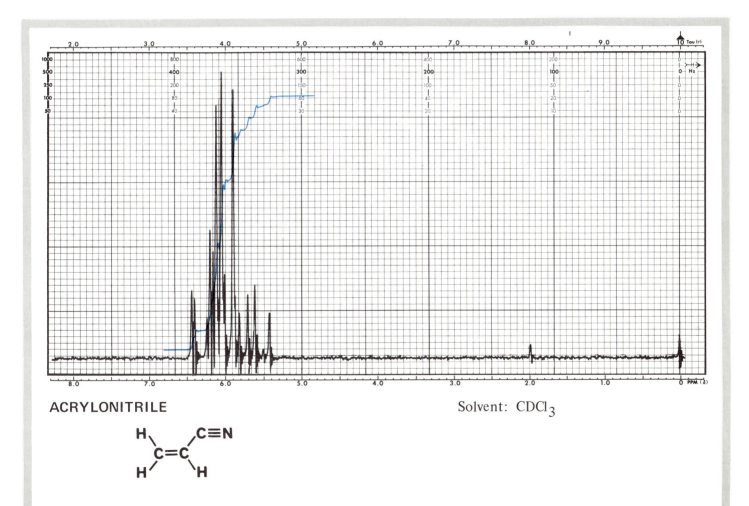

ACRYLONITRILE

Solvent: CDCl$_3$

197

The two hydrogens of vinylidene chloride are equivalent and resonate as a single sharp peak at 5.51 ppm. Upon polymerization, these olefinic hydrogens become an aliphatic CH_2 group which resonates at much higher field, near 3.0 ppm. The monomer is often copolymerized with vinyl chloride or acrylonitrile to produce plastics which are highly inert to chemical attack.

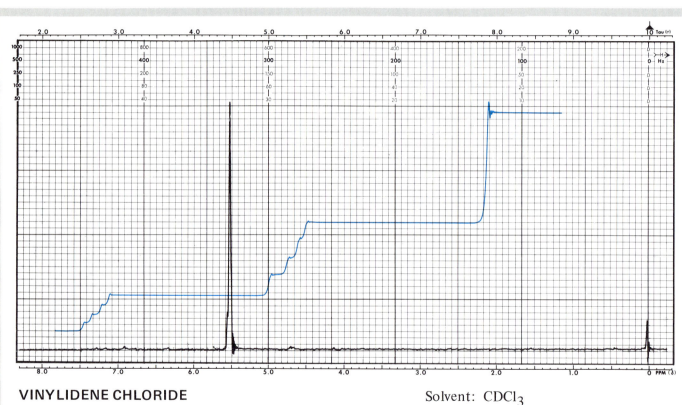

VINYLIDENE CHLORIDE

Solvent: $CDCl_3$

The acetate methyl group is observed at highest field as a sharp three proton singlet. The three vinyl protons resonate at lower field. The proton which is alpha to the carboxylate group resonates at 7.25 and the two protons on the beta carbon are centered near 4.7 ppm. This olefinic proton pattern is quite different from that observed for the vinyl protons of the acrylate monomers.

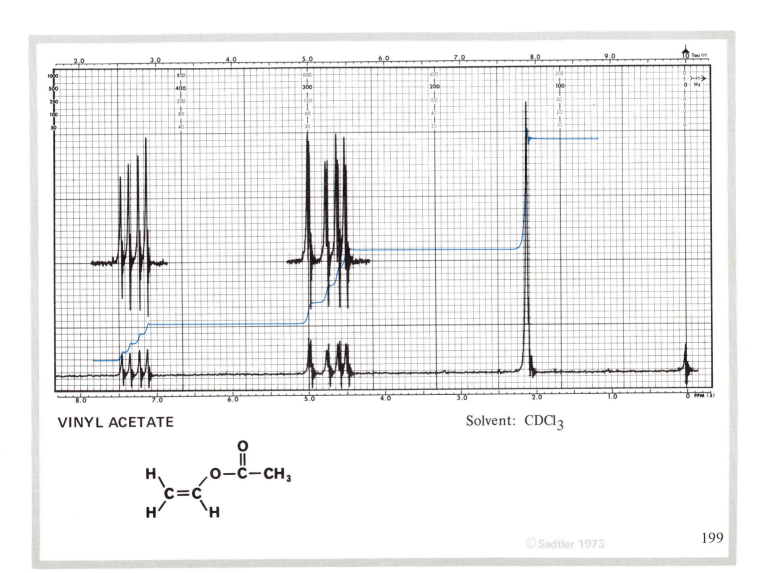

VINYL ACETATE

Solvent: $CDCl_3$

199

The three vinyl protons of acrylic acid form a complex higher-order pattern in the chemical shift range from 5.7 to 6.9 ppm. The carboxylic acid proton which resonates at about 12.3 ppm was offset to higher field and recorded near the left margin of the chart.

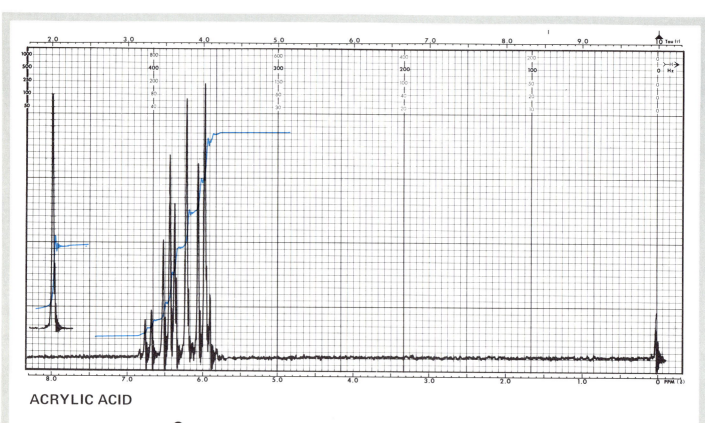

ACRYLIC ACID

The strongest band in the spectrum of methyl acrylate is the methyl ester group which resonates as a sharp singlet at 3.78 ppm. The three vinyl protons appear as a complex higher order pattern in the chemical shift range from 5.6 to 6.6 ppm. With the loss of the carbon-carbon double bond during polymerization, the methyl group undergoes a slight shift to higher field and resonates at about 3.6 ppm.

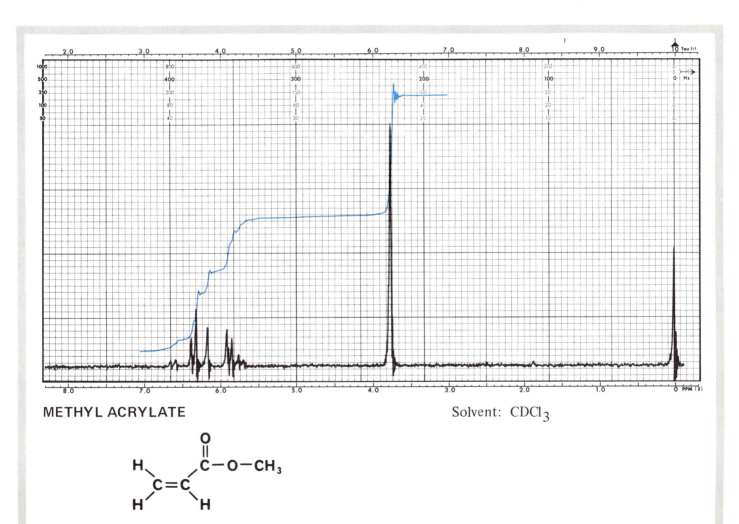

METHYL ACRYLATE

Solvent: $CDCl_3$

201

The methyl and methylene groups resonate at 1.3 ppm and 4.2 ppm as a triplet and quartet respectively. The three olefinic protons form a complex higher order pattern in the range from 5.6 to 6.7 ppm. Upon polymerization, the olefinic protons are converted into aliphatic methylene and methine groups and appear at much higher field (0.8 to 3.0 ppm).

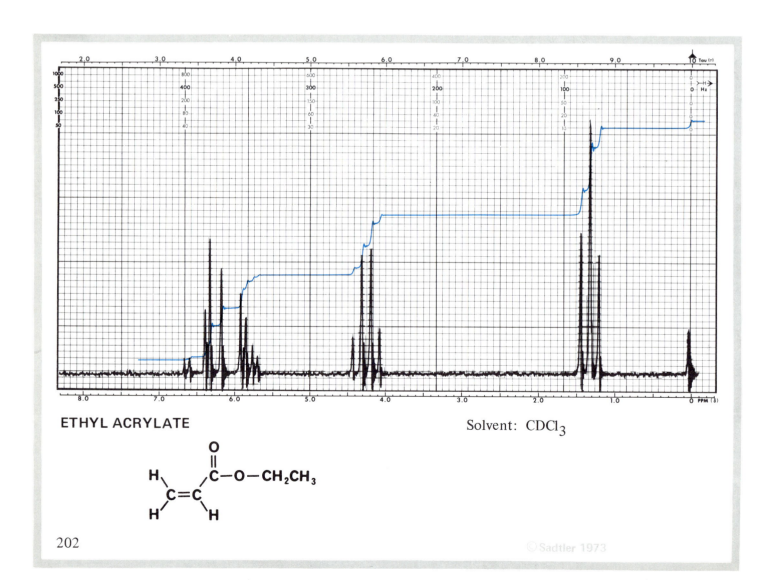

ETHYL ACRYLATE

Solvent: CDCl$_3$

The three vinyl protons of this monomer produce a pattern similar to that observed for acrylic acid and methyl acrylate. The terminal methyl group of the butyl moiety resonates as a broadened triplet at 0.97 ppm; the two adjacent methylene groups form a complex multiplet centered at about 1.5 ppm, and the CH_2 group which is bonded to the oxygen atom appears at 4.2 ppm as a clear triplet.

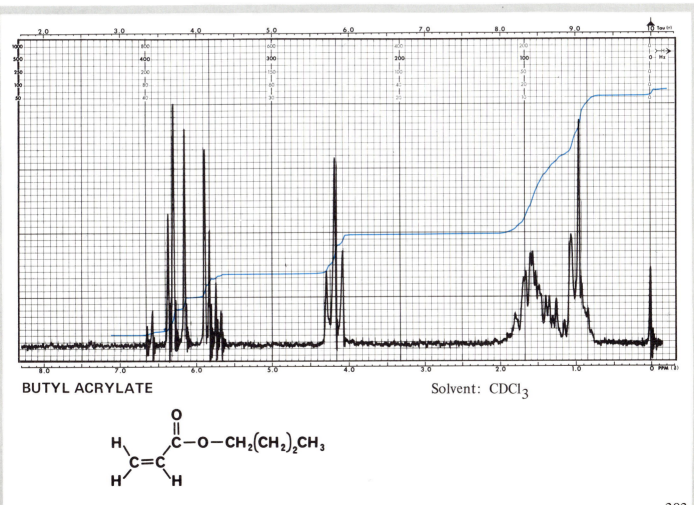

BUTYL ACRYLATE

Solvent: $CDCl_3$

203

The characteristic higher order pattern for the three vinyl protons is observed at low field (5.6 to 6.7 ppm). The isobutyl structure produces a dramatically different series of patterns than did the n-butyl moiety (page 203).   As predicted by the N + 1 rule, the terminal methyl groups resonate as a six proton doublet at highest field. The methine is observed at 2.00 ppm. The methylene group bonded to the ester oxygen atom appears at 3.95 ppm as a clear doublet.

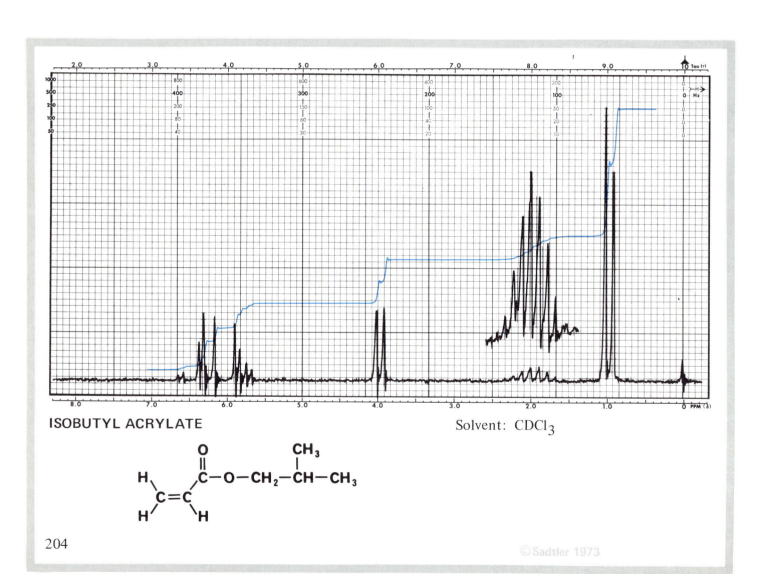

ISOBUTYL ACRYLATE

Solvent: $CDCl_3$

The methacrylate methyl group absorbs at 1.95 ppm and displays long range coupling to the two vinyl protons. The methyl ester group appears as a sharp singlet at 3.75 ppm. The two non-equivalent olefinic hydrogens possess chemical shifts of 5.55 and 6.11 ppm respectively and display long range coupling to the methyl group on the adjacent carbon atom.

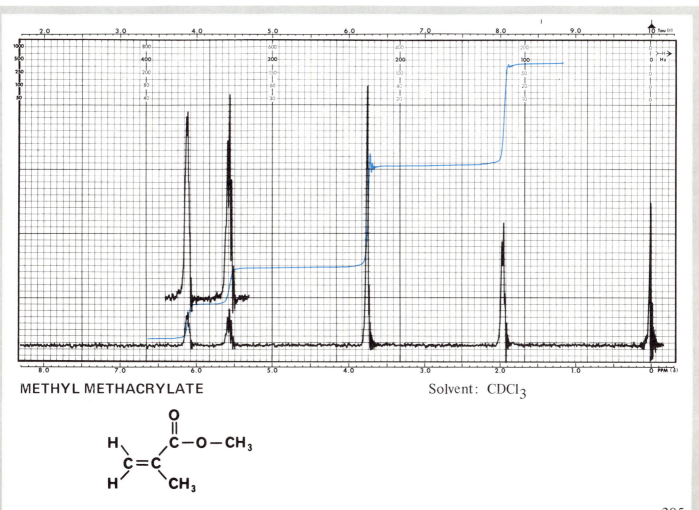

METHYL METHACRYLATE

Solvent: $CDCl_3$

205

Ethyl methacrylate displays the characteristic triplet and quartet pattern for the ethyl ester group. The methyl triplet resonates at 1.30 ppm and the quartet appears at about 4.21 ppm. The methacrylate methyl group is observed at 1.95 ppm as a broad single peak displaying long range coupling to the nearby olefinic hydrogens. The two vinyl protons are non-equivalent and resonate at 5.53 and 6.11 ppm respectively. Upon polymerization, the chemical shifts of the triplet and quartet are only slightly affected while those of the methacrylate protons undergo a large shift to higher field (see page 66).

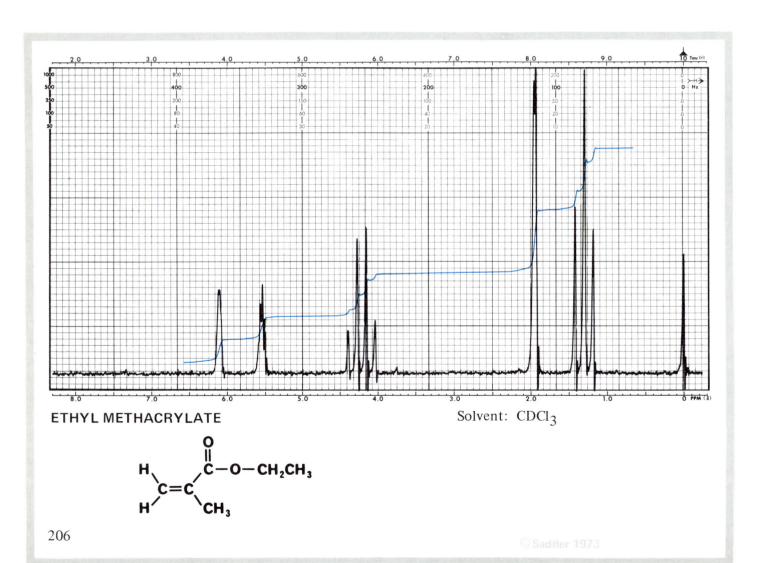

ETHYL METHACRYLATE

Solvent: CDCl$_3$

This spectrum of the monomer of polyvinyl pyrrolidone contains four distinct sets of absorption bands. At highest field, the methylene groups which are alpha and beta to the carbonyl group overlap to form a complex multiplet in the chemical shift range from 1.8 to 2.7 ppm. The methylene group alpha to the nitrogen atom appears at 3.51 as a clear triplet. The two terminal olefinic protons occur as overlapping doublets centered near 4.4 ppm. The methine which is alpha to the nitrogen atom resonates at lowest field as a cis/trans doublet of doublets at 7.1 ppm.

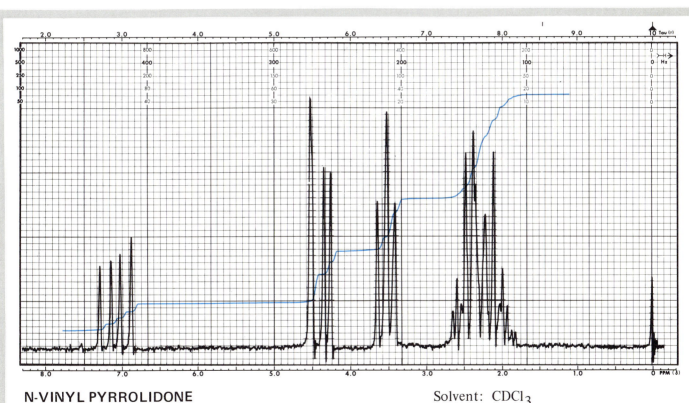

N-VINYL PYRROLIDONE

Solvent: $CDCl_3$

$H_2C=CH-N$

207

Ethylene glycol (and its cyclic analog-ethylene oxide) are used as monomers in the preparation of poly-ethers (p. 88 , p. 94 , p. 163) and polyesters (p. 101, p. 103). In this spectrum, the two equivalent methylene groups of the monomer absorb as a sharp single peak at 3.7 ppm. This chemical shift is similar to that observed for the methylene groups of its polyether. When ethylene glycol is incorporated into a polyester, however, the methylene groups undergo a shift to lower field and absorb in the range from 4.5 to 5.0 ppm.

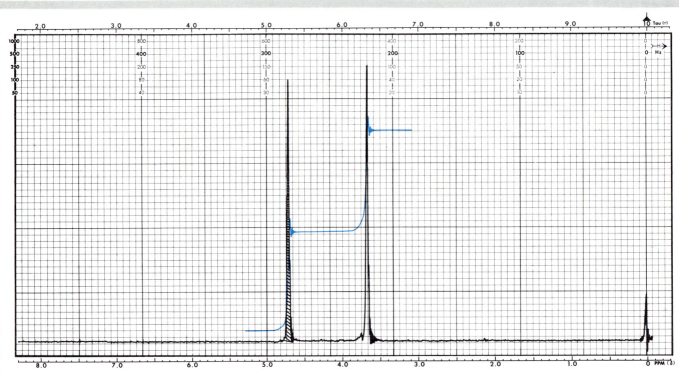

ETHYLENE GLYCOL   Solvent: D_2O

$$HO-CH_2CH_2-OH$$

208

The chemical shifts of the CH$_2$ groups adjacent to the ether oxygen atoms are very similar to those of the CH$_2$ groups which are adjacent to the hydroxyl oxygens, resulting in a complex higher order A$_2$B$_2$ pattern centered at about 3.7 ppm. When incorporated into a polyester, the methylene groups adjacent to the carboxylic acid oxygen atoms are selectively deshielded resulting in the appearance of two separate multiplets (see page 102).

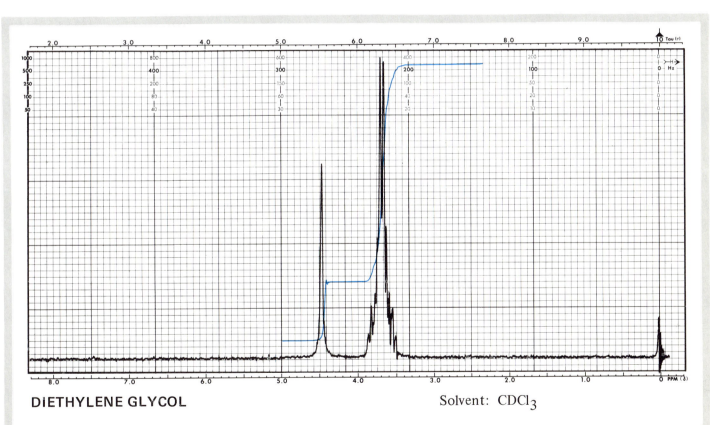

DIETHYLENE GLYCOL

Solvent: CDCl$_3$

$$HO - CH_2 - CH_2 - O - CH_2 - CH_2 - OH$$

209

The methyl group appears at highest field as a three proton doublet near 1.17 ppm. The methylene protons resonate at 3.5 ppm and the methine appears as a complex multiplet at 3.86 ppm. The single peak at 4.69 represents the two hydroxyl groups in exchange with the solvent D_2O. The methylene group does not appear as a doublet because its two protons are not equivalent.

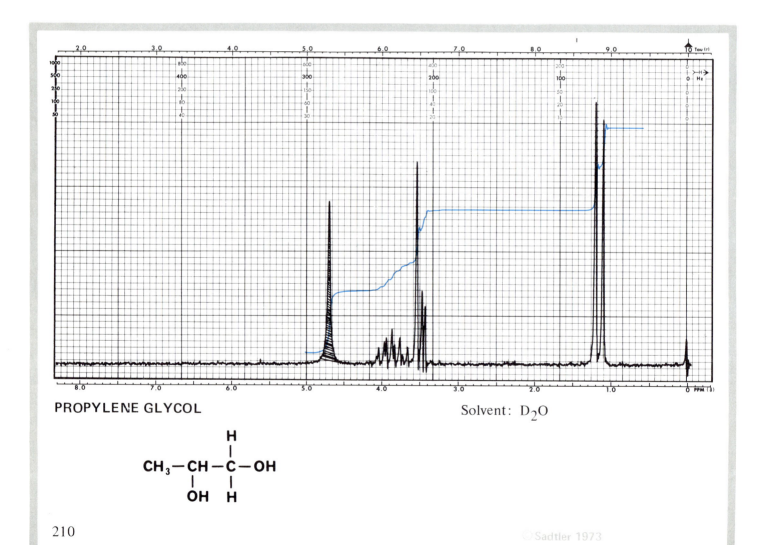

PROPYLENE GLYCOL

Solvent: D_2O

210

The two centrally situated methylene groups appear at highest field as a distorted triplet while those which are bonded to the hydroxyl groups are observed near 3.62 ppm. The characteristic appearance of these multiplets results from virtual coupling. The hydroxyl groups are in exchange with the deuterium atoms of the solvent resulting in the HDO band at 4.7 ppm. 1,4-Butanediol is a common constituent of polyesters and polyurethanes (see p. 104).

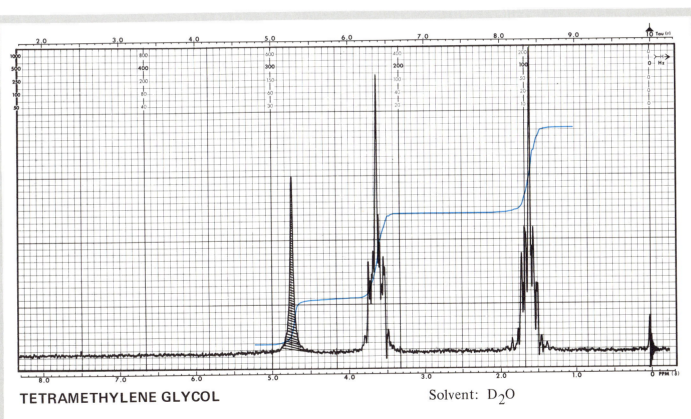

TETRAMETHYLENE GLYCOL

Solvent: D_2O

$$HO-CH_2(CH_2)_2CH_2-OH$$

211

The four different hydrocarbon groups of 1,3-butylene glycol produce a distinct set of multiplets including a quartet at unusually high field which represents the methylene group in the center of the molecule. It is coupled to the adjacent methine and methylene groups. At highest field, the methyl protons resonate as the expected three proton doublet. At lower field, the methylene group adjacent to the terminating hydroxyl group appears as a triplet centered at 3.71 ppm overlapping with the methine pattern which is observed as a sextet centered at 3.9 ppm. The two hydroxyl protons are in rapid exchange and appear as a single two proton peak at 4.34 ppm.

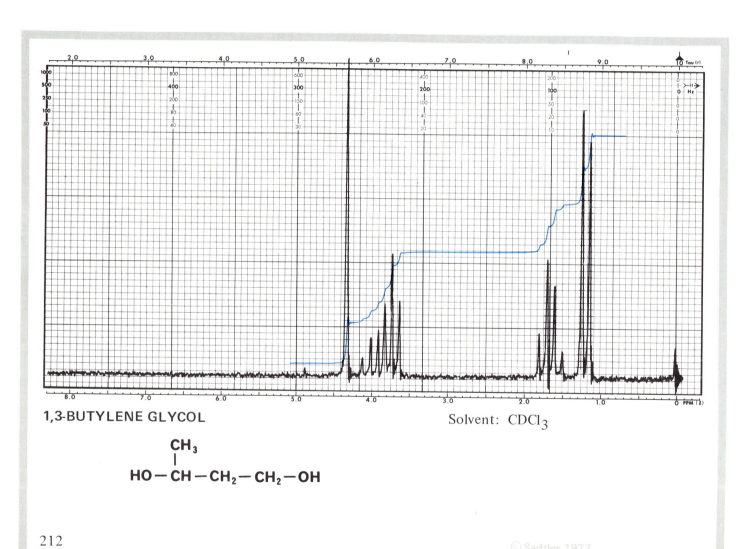

1,3-BUTYLENE GLYCOL

Solvent: $CDCl_3$

$$HO-\underset{\underset{CH_3}{|}}{CH}-CH_2-CH_2-OH$$

212

The one methine and the two methylene groups of glycerol, although non-equivalent possess similar chemical shifts and produce a higher-order multiplet centered at 3.63 ppm. The three hydroxyl protons have been replaced by deuterium from the solvent with the formation of HDO as a solvent band which resonates at 4.71 ppm. Upon esterification, the methylene and methine groups will possess significantly different chemical shifts producing a very different absorbance pattern.

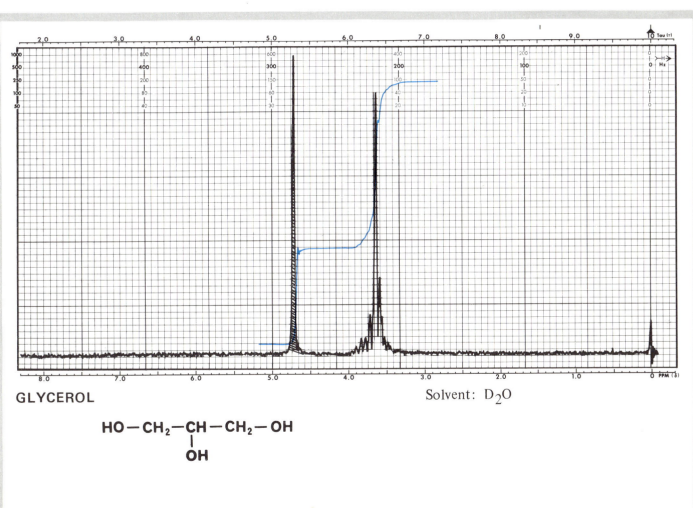

GLYCEROL

Solvent: D_2O

$$HO-CH_2-CH-CH_2-OH$$
$$OH$$

213

1,2,6-Hexanetriol is an intermediate in the preparation of polyethers and polyesters where the third hydroxyl group provides a site at which cross-linking of the polymer chains may occur. The four CH_2 groups in the center of the chain resonate as a broad band at highest field near 1.45 ppm. The methine and methylenes which are substituted by the hydroxyl groups are deshielded to a similar degree and overlap to form a complex multiplet at about 3.52 ppm. The hydroxyl protons are in exchange with the solvent and contribute to the intensity of the HDO band at 4.69 ppm.

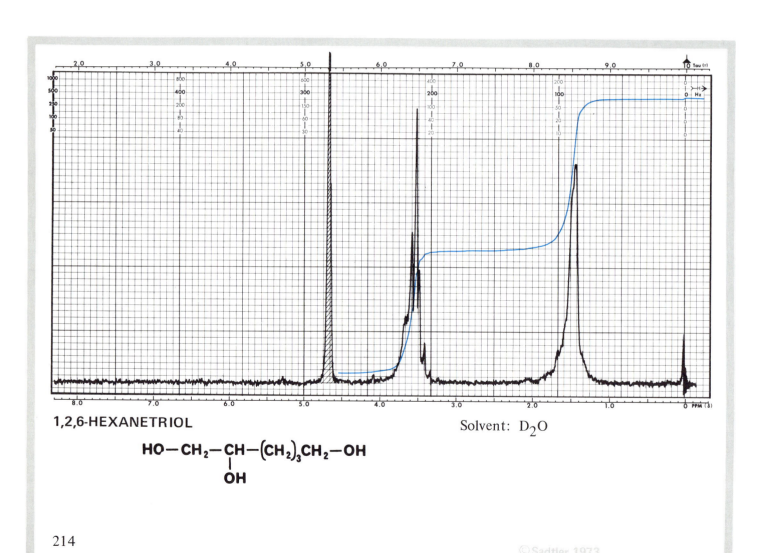

1,2,6-HEXANETRIOL

Solvent: D_2O

$$HO-CH_2-CH-(CH_2)_3CH_2-OH$$
$$|$$
$$OH$$

The three furan ring hydrogens resonate at lowest field. H-3 and H-4 overlap at 6.28 ppm while H-5 appears at 7.35 due to the strong deshielding effect of the adjacent oxygen atom. The methylene group appears as a sharp singlet at 4.5 ppm and the hydroxyl proton resonates at 3.49 ppm. Very insoluble thermosetting resins are prepared from furfuryl alcohol.

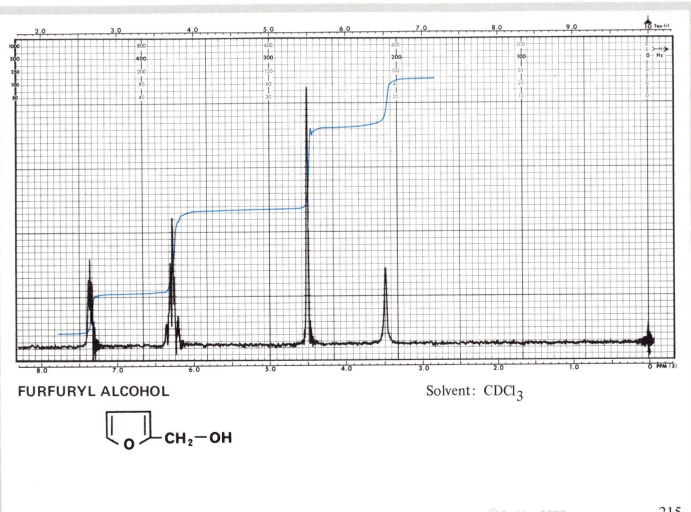

FURFURYL ALCOHOL

Solvent: $CDCl_3$

The hydroxyl substituent on the aromatic ring preferentially shields the ortho and para ring hydrogens producing a complex pattern of absorbance bands in the chemical shift range from 6.7 to 7.4 ppm. The phenolic proton resonates as a sharp singlet at 5.9 ppm. Reactions involving the replacement of the phenolic hydrogen or additional substitution of the aromatic ring usually result in a deshielding of the aromatic absorbance bands.

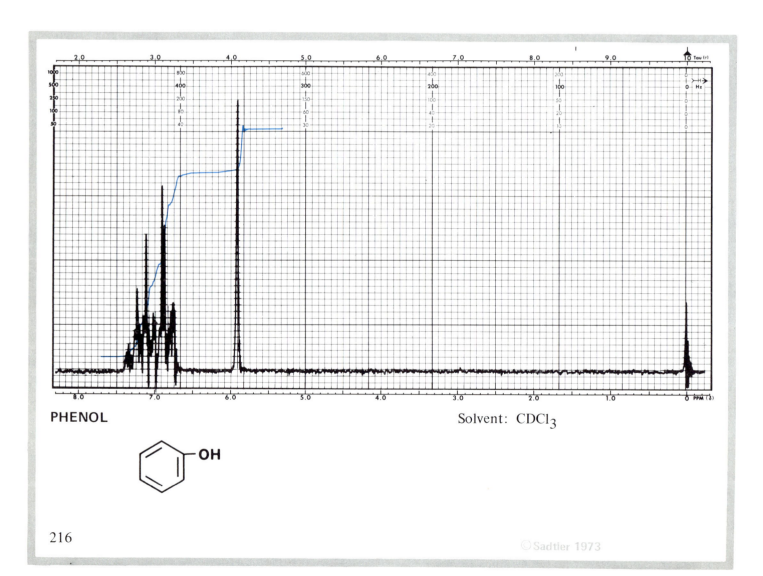

PHENOL

Solvent: CDCl$_3$

Para tert-butyl phenol is a monomer for certain phenol-formaldehyde resins such as those presented on pages 152 and 153. The t-butyl group bonded to the benzene ring resonates as a very sharp nine proton singlet at 1.28 ppm. The sharpness of this peak and the absence of a terminal methyl band at 0.89 ppm help to distinguish a t-butyl group such as this from the fatty acid methylene band which appears at about the same chemical shift (compare with stearic acid on page 237). The aromatic protons display a clear para disubstituted benzene pattern at lowest field. The hydrogens ortho to the hydroxyl group resonate at 6.71 and those ortho to the t-butyl group appear at 7.19 ppm. The exchangeable phenolic proton is observed as a broad band at about 5.19 ppm.

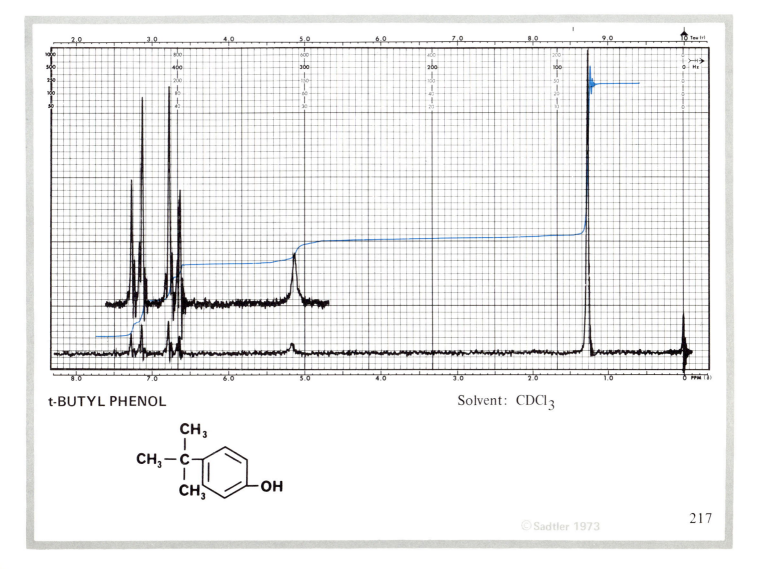

t-BUTYL PHENOL

Solvent: $CDCl_3$

©Sadtler 1973

The sharp singlet at 1.6 ppm represents the two equivalent methyl groups. At lower field, there is observed the characteristic para substitution ortho doublets (6.74, 7.06) for the eight aromatic protons. At lowest field, the phenolic protons resonate at 7.89 ppm in this spectrum. Except for the loss of the phenolic proton absorbance band, the chemical shifts of bisphenol-A are not strongly affected by the formation of an aliphatic ether linkage.

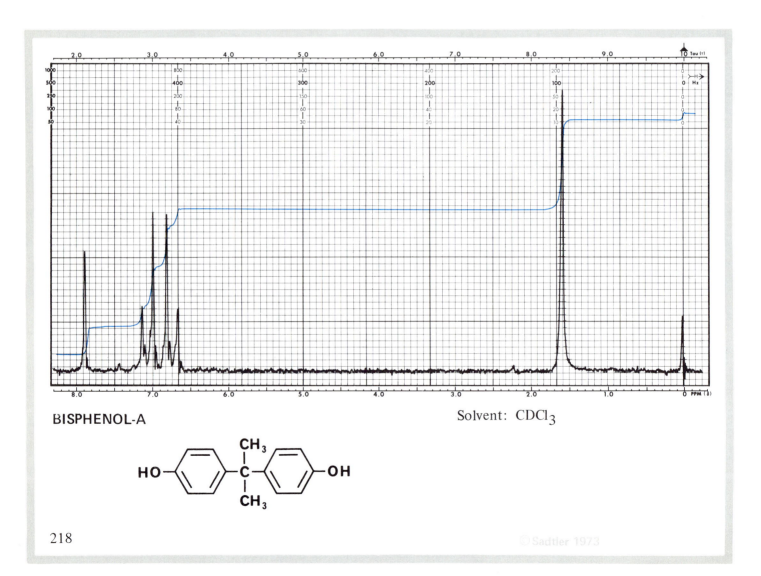

BISPHENOL-A

Solvent: $CDCl_3$

This monomer is used in the formation of poly(phenylene oxide) (p. 95). The two methyl groups which are bonded to the aromatic ring resonate at 2.2 ppm. The three aromatic protons appear in the chemical shift range from 6.6 to 7.1 ppm as a complex higher order pattern. In this spectrum, the phenolic proton appears as a sharp singlet at 4.6 ppm.

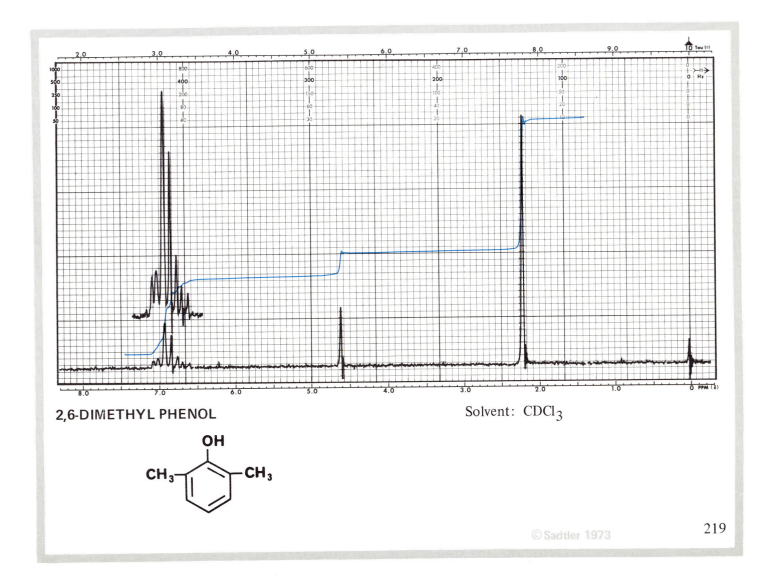

2,6-DIMETHYL PHENOL

Solvent: $CDCl_3$

219

Dihydroxy diphenyl sulfone is a monomer for polysulfones such as that appearing on page 96 . The aromatic hydrogens which are ortho to the hydroxyl groups resonate as an ortho doublet at 6.87 ppm. The protons which are ortho to the sulfonyl group appear as an ortho doublet at 7.67 ppm. The two phenolic protons were not recorded on the spectrum but were observed as a very broad band with a chemical shift of about 10.5 ppm. The band at 3.67 ppm is due to the presence of a small amount of water present in the sample solution.

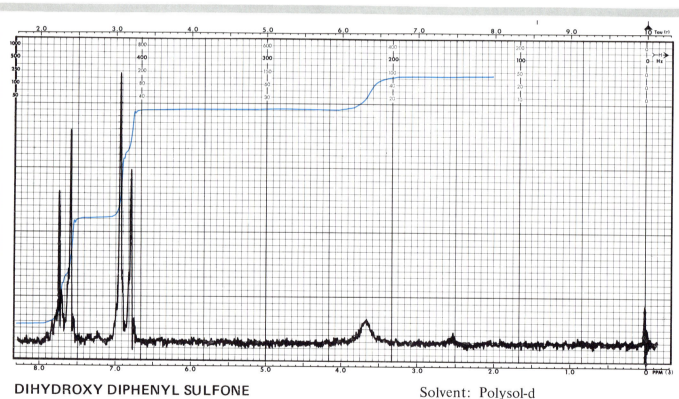

DIHYDROXY DIPHENYL SULFONE

Solvent: Polysol-d

The proton spectrum of epichlorohydrin is quite complex due to the fact that all five hydrogens in the molecule are non-equivalent yet resonate within a relatively narrow chemical shift range (2.5 to 4 ppm). The two non-equivalent protons of the methylene chloride group appear at lowest field (3.4 to 3.8 ppm). At 3.17, the epoxy methine occurs as a very complex multiplet. The two non-equivalent ring methylene protons resonate at 2.84 and 2.6 ppm. The first step in many polymer reactions involving epichlorohydrin is the loss of chlorine and the formation of an ether linkage, followed by opening of the epoxide ring to form a glycerine-like chain.

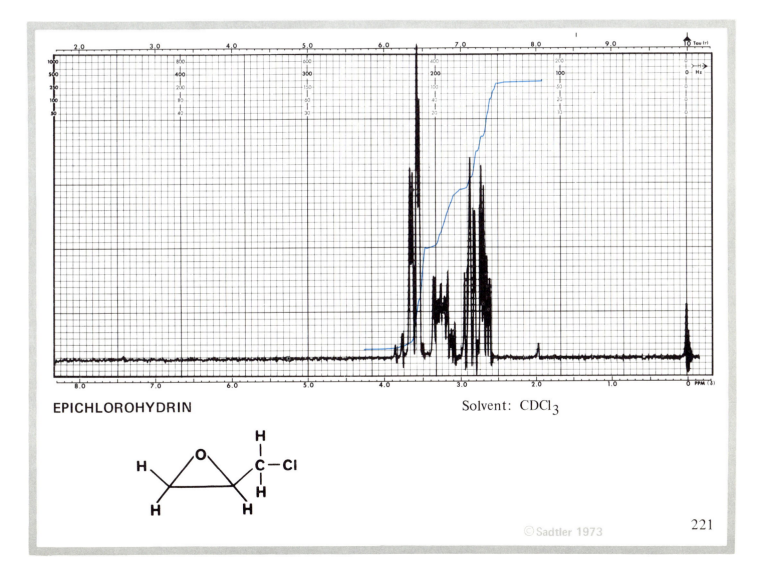

EPICHLOROHYDRIN

Solvent: CDCl$_3$

221

The eight cyclohexane methylene protons resonate at highest field in the chemical shift range from 1.1 to 2.4 ppm. The ring methines adjacent to the ether oxygen substituents are not clearly observed but one is probably represented by the weak multiplet at 3.63 ppm. The protons of the epoxide ring system resonate as a complex series of multiplets in the chemical shift range from 2.5 to 3.4 ppm. The glycidyl methylene groups appear as a series of doublets in the range from 3.7 to 4.6 ppm. The protons of each CH_2 group are non-equivalent and show a much larger difference in chemical shift than is observed for the same protons of epichlorohydrin on the previous page.

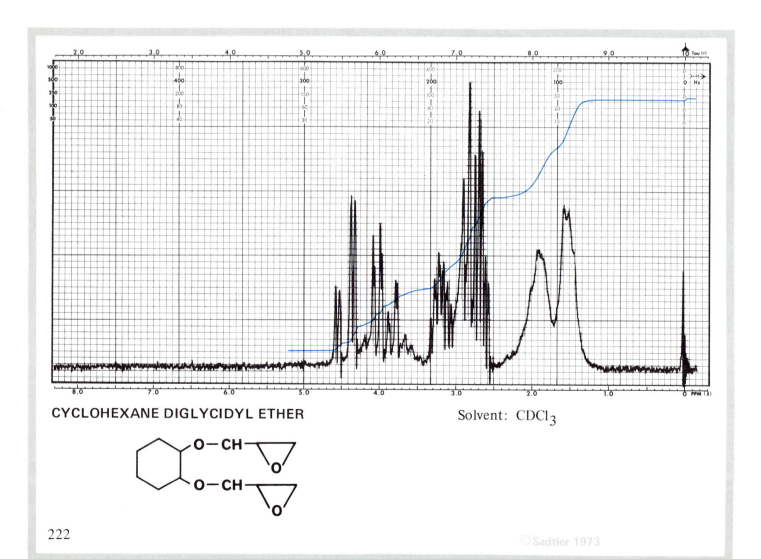

CYCLOHEXANE DIGLYCIDYL ETHER

Solvent: $CDCl_3$

1,6-Hexanediisocyanate produces an NMR spectrogram consisting of two absorbance bands. The four methylene groups in the center of the chain overlap to form a broad band centered at about 1.43 ppm. The CH$_2$ groups adjacent to the isocyanate groups appear as a broadened triplet at 3.29 ppm. Some polymerization of the diisocyanate monomer is indicated by the deshielding of some of the methylene groups from 3.29 to form a new band at 3.69 ppm and the appearance of an exchangeable amide proton band at 7.28 ppm.

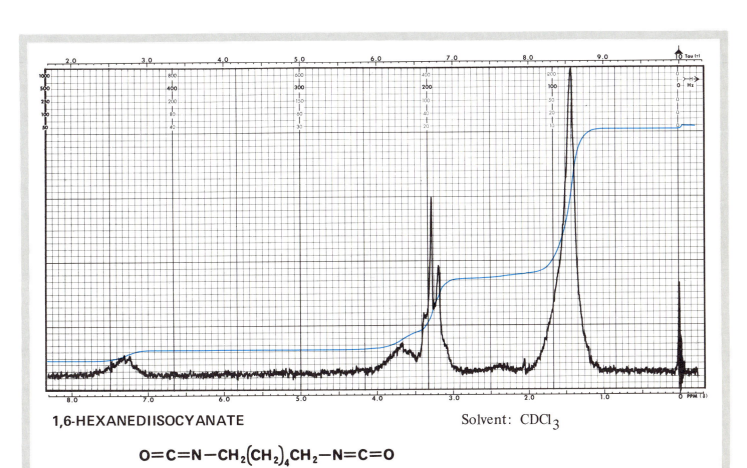

1,6-HEXANEDIISOCYANATE

Solvent: CDCl$_3$

$$O=C=N-CH_2(CH_2)_4CH_2-N=C=O$$

The commercial tolylene diisocyanate which is used in the preparation of urethane polymers is often a mixture of the 2,4- and 2,6-diisocyanate isomers. The NMR spectrogram on this page is that of a sample which contains about 80% of 2,4-tolylene diisocyanate and about 20% of the 2,6- isomer. The methyl groups of both forms are accidentally equivalent and resonate as a single peak at 2.29 ppm. The high field (above 7.0 ppm) chemical shifts of the majority of the aromatic absorbance bands reflects the strong shielding effect of isocyanate groups on ortho and para benzene protons. The narrow peak at 6.93 ppm represents the center peak of the di-ortho triplet of the 2,6- component (compare with the spectrum on the next page).

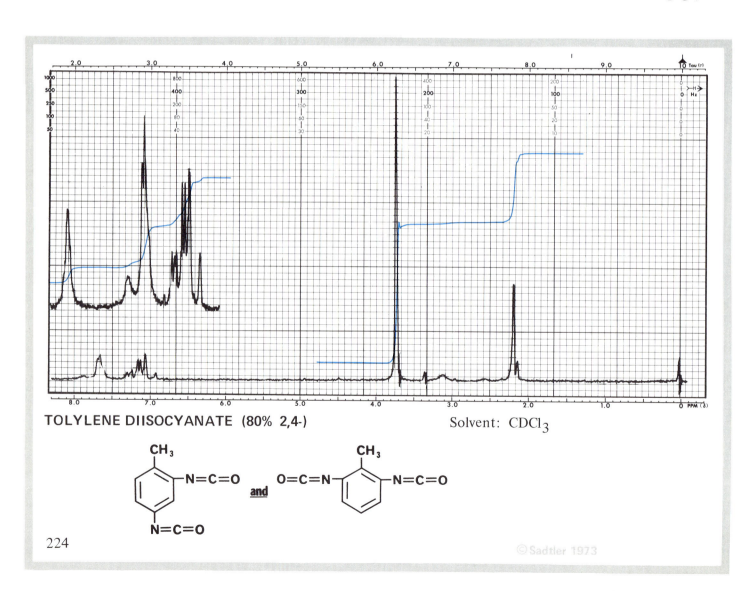

TOLYLENE DIISOCYANATE (80% 2,4-)   Solvent: CDCl$_3$

This spectrum represents the reaction product of the TDI sample on page 224 with methanol. It illustrates the chemical shift changes which occur in the aromatic region upon the conversion of the isocyanate groups into carbamic acid ester groups. The aromatic resonance bands have all undergone a significant shift to lower field in comparison to those of the TDI monomer.

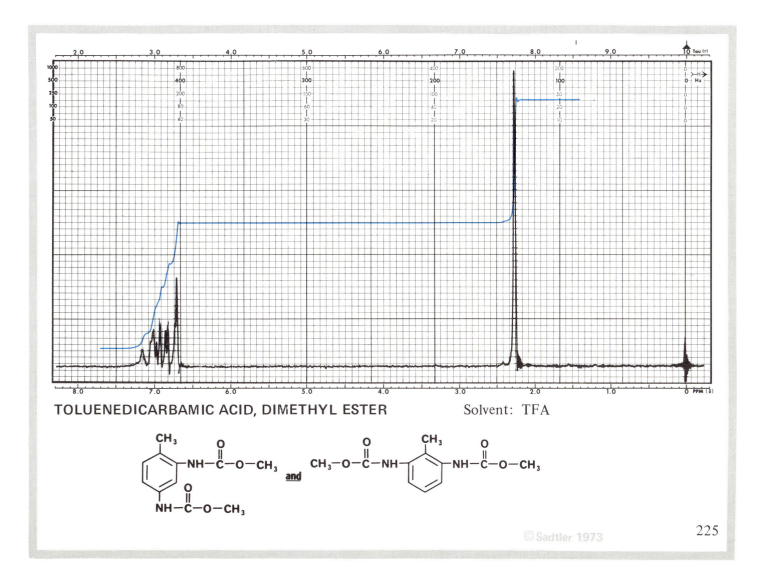

TOLUENEDICARBAMIC ACID, DIMETHYL ESTER   Solvent: TFA

225

Although similar in general appearance to the sample of tolylene diisocyanate depicted on page 224, this material contains only 65% of the 2,4- isomer and 35% of the 2,6-tolylene diisocyanate. The two NMR spectra display slight differences in the intensity of the bands in the aromatic region. The band at 6.97 ppm in this spectrum is taller than that of the earlier spectrum while the band at 6.75 ppm is diminished in height.

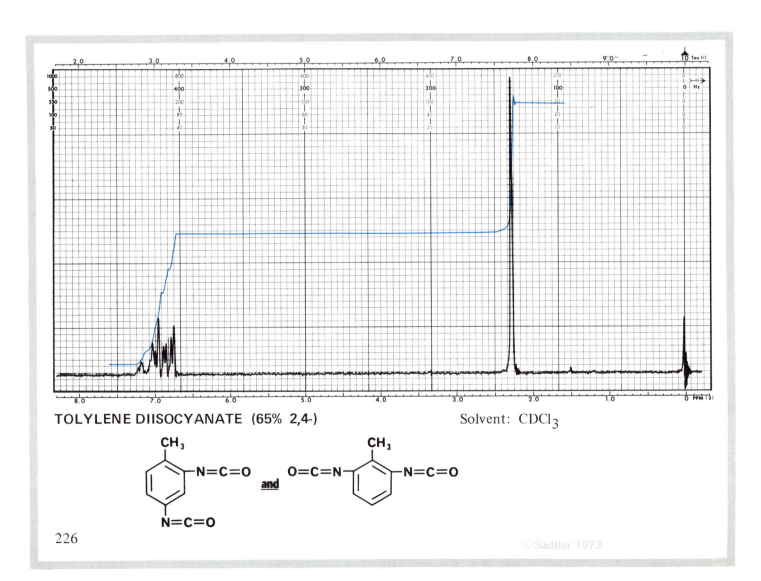

TOLYLENE DIISOCYANATE (65% 2,4-)   Solvent: $CDCl_3$

Tolidine diisocyanate (TODI; 3,3'-dimethyl-4,4'-biphenyl diisocyanate) is a commonly utilized monomer for the preparation of urethane polymers. The two equivalent methyl groups resonate at 2.38 ppm as a sharp singlet. The hydrogen adjacent to the isocyanate group (H-5) appears as a clear ortho doublet at 7.11 ppm while the hydrogen adjacent to the biphenyl linkage (H-5) resonates at 7.32 ppm as an ortho/meta doublet of doublets. The hydrogen ortho to the methyl group (H-2) resonates at lowest field as a meta doublet.

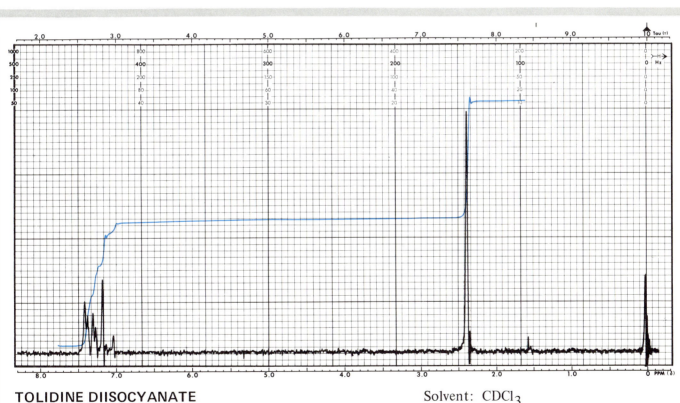

TOLIDINE DIISOCYANATE

Solvent: CDCl$_3$

227

MDI (methylene-diparaphenylene isocyanate) is used in the preparation of polyurethane resins and spandex fibers, in which it reacts with alcohols to form urethane linkages. In this NMR spectrum, the methylene bridge between the two aromatic rings resonates at 3.92 ppm. The four hydrogens of each aromatic ring appear as a higher order para substitution pattern centered at 7.09 ppm. The weak singlet at 1.51 ppm is due to an impurity in the sample solution.

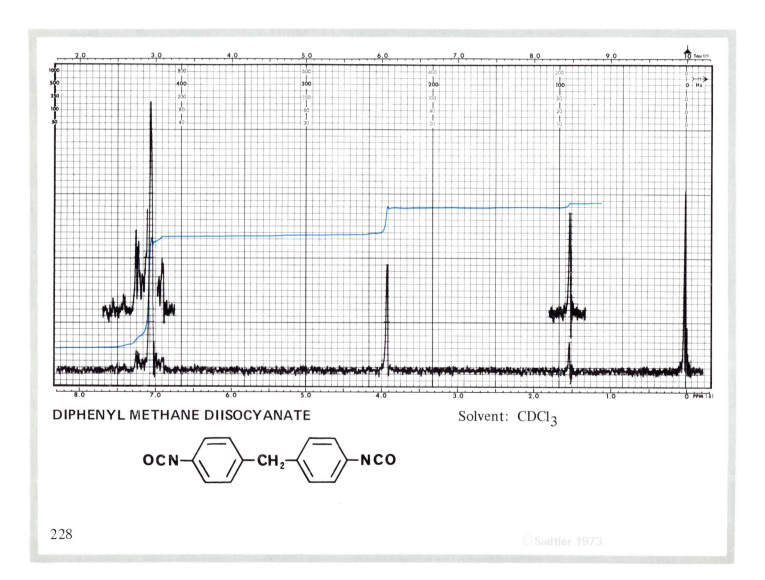

DIPHENYL METHANE DIISOCYANATE

Solvent: $CDCl_3$

Adipic acid produces an NMR spectrogram consisting of three absorption bands. The two methylene groups in the center of the molecule resonate at highest field near 1.61 ppm as a four proton distorted triplet. The methylene groups bonded to the carbonyl groups resonate at lower field near 2.25 ppm. The carboxylic acid protons appear at much lower field and in this spectrum resonate at 10.33 ppm. The two characteristic distorted triplets at high field are easily recognized in polymers by their characteristic shape and chemical shifts.

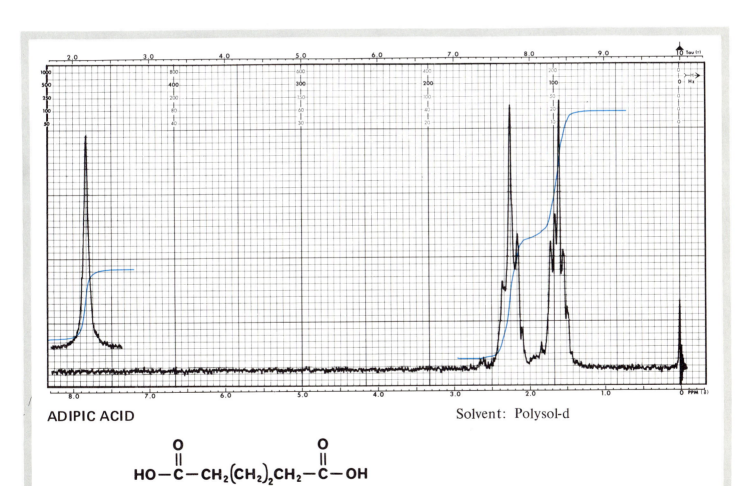

ADIPIC ACID

Solvent: Polysol-d

$$HO-\overset{\overset{\textstyle O}{\|}}{C}-CH_2(CH_2)_2CH_2-\overset{\overset{\textstyle O}{\|}}{C}-OH$$

229

The five CH$_2$ groups in the center of the molecule resonate at highest field as a broad band in the chemical shift range from 1.1 to 1.9 ppm. The strongest band in the spectrum is located at 1.37 ppm. The methylene groups which are bonded to the carbonyl carbons resonate at 2.21 ppm as a broadened triplet. The carboxylic acid protons appear in the offset range below 8.3 ppm and possess a chemical shift of about 8.91 ppm. The integration value of this band indicates that a small amount of water from the solvent is in exchange with the carboxylic acid protons.

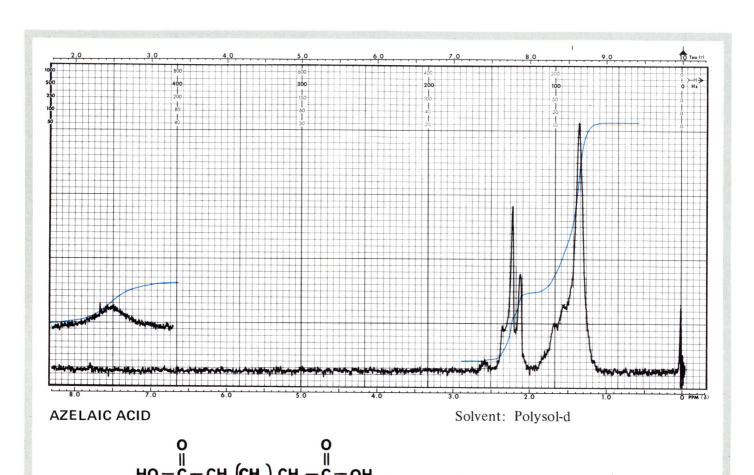

AZELAIC ACID

Solvent: Polysol-d

$$HO-\overset{\overset{\displaystyle O}{\|}}{C}-CH_2(CH_2)_5CH_2-\overset{\overset{\displaystyle O}{\|}}{C}-OH$$

230

Although visually similar to the spectrum of azelaic acid on the previous page, the NMR spectrogram of sebacic acid differs in the integration value of the high field absorbance band in the chemical shift range from 1.1 to 1.9 ppm. In addition, the strongest peak in the spectrum appears at 1.31 ppm instead of 1.37 ppm as a result of the additional methylene group present in the sebacic acid chain. The carboxylic acid protons resonate in the offset range with a chemical shift of about 11.28 ppm.

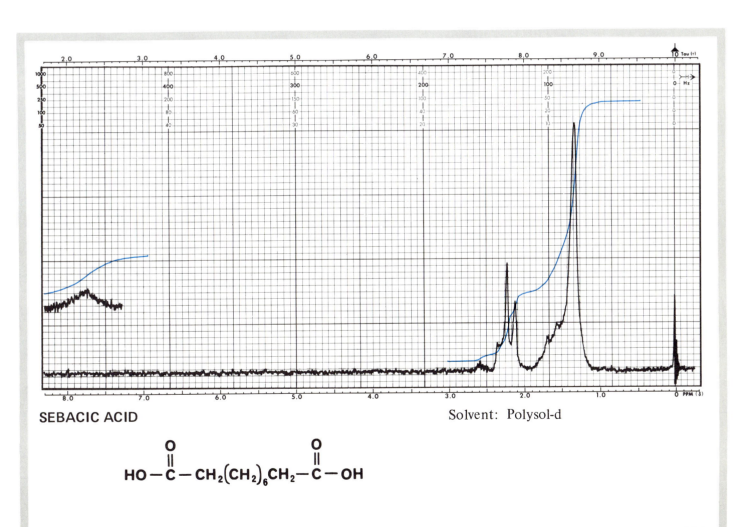

SEBACIC ACID

Solvent: Polysol-d

$$HO-\underset{\underset{O}{\|}}{C}-CH_2(CH_2)_6CH_2-\underset{\underset{O}{\|}}{C}-OH$$

231

Isophthalic acid produces a characteristic low field, meta disubstituted benzene pattern. The four lines at 7.67 ppm represent the resonance of H-5, a di-ortho higher order triplet. The ortho/di-meta doublet of triplets centered at 8.21 represents H-4 and H-6 and the di-meta triplet at 8.58 ppm represents the resonance of H-2. The carboxylic acid protons appear in the spectrum as a very broad band centered in the offset range at about 12.25 ppm. The five line multiplet at about 2.6 ppm arises from the solvent impurity DMSO-d5.

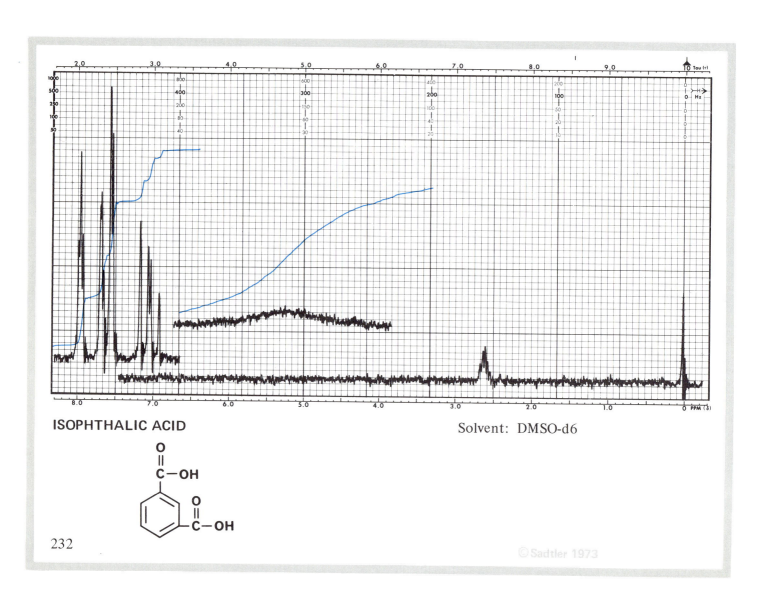

ISOPHTHALIC ACID

Solvent: DMSO-d6

232

The four equivalent aromatic protons of terephthalic acid resonate at 8.08 ppm as a sharp singlet. The presence of water in this sample solution has significantly increased the integration value of the hydroxyl protons and has resulted in a shift of their resonance to higher field (9.0 ppm) from a "normal" benzoic acid resonance of about 12.0 ppm. The sharp single peak at low field is readily discernable in a number of polyesters such as those appearing on pages 108, 109, 111 and 112.

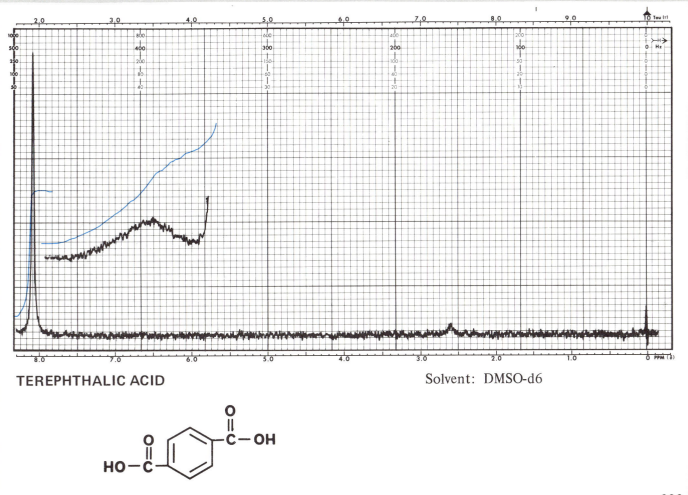

TEREPHTHALIC ACID

Solvent: DMSO-d6

233

Lauric acid, the eleven carbon saturated fatty acid, occurs as a triglyceride in many natural fats and oils. Its NMR spectrogram contains a broadened triplet at highest field (0.89 ppm) which represents the terminal methyl groups. The nine CH_2 groups in the center of the chain appear as a single sharp peak at about 1.28 ppm while the methylenes adjacent to carbonyl groups resonate as a broadened triplet at about 2.31 ppm. The carboxylic acid proton resonates as a sharp singlet at 11.62 ppm.

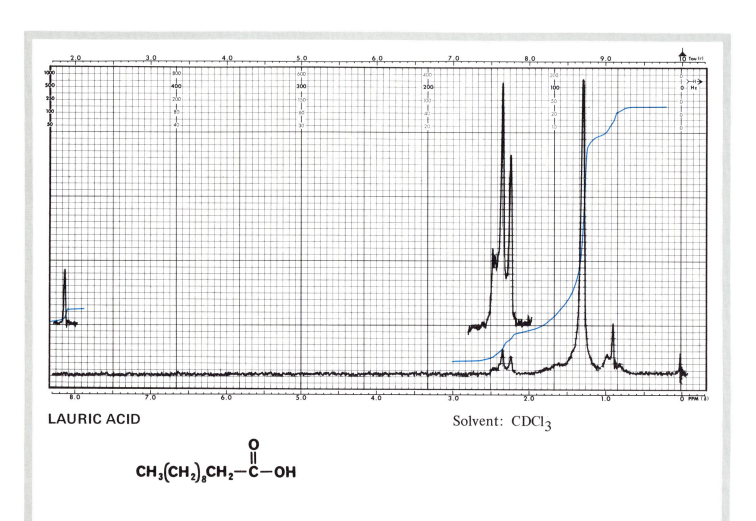

LAURIC ACID

Solvent: $CDCl_3$

$$CH_3(CH_2)_8CH_2-\overset{\overset{\displaystyle O}{\|}}{C}-OH$$

234

Myristic acid, the fourteen carbon saturated fatty acid, produces a spectrum which is similar in appearance and chemical shifts to that of lauric acid on the previous page. The absorbance bands can be assigned as follows:

CH_3-	triplet at 0.89 ppm
$-(CH_2)_{11}-$	1.1 to 1.9 ppm
$-CH_2-C(=O)-$	triplet at 2.31 ppm
$-C(=O)-OH$	singlet at 11.53 ppm

This spectrum differs from those of the other fatty acids only in the integration value of the methylene band in the chemical range from 1.1 to 1.9 ppm. For myristic acid, the integration value in this range should indicate the presence of 22 hydrogens (eleven CH_2 groups).

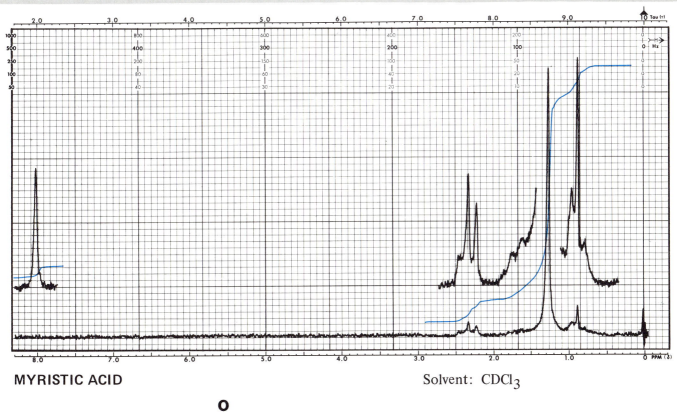

MYRISTIC ACID

Solvent: $CDCl_3$

$$CH_3(CH_2)_{11}CH_2-\overset{\overset{\textstyle O}{\|}}{C}-OH$$

235

Palmitic acid, which is the sixteen carbon saturated fatty acid, displays an NMR spectrum similar in general appearance to those of the three previous fatty acids. The distinguishing feature is that the integration value of the methylene proton band in the chemical shift range from 1.1 to 2.5 ppm indicates the presence of a total of fourteen CH_2 groups (28 protons). The complex multiplet which is centered at about 1.6 ppm represents the methylene group which is beta to the carbonyl. The strongest band in the spectrum is the 26 proton single peak at 1.28 ppm.

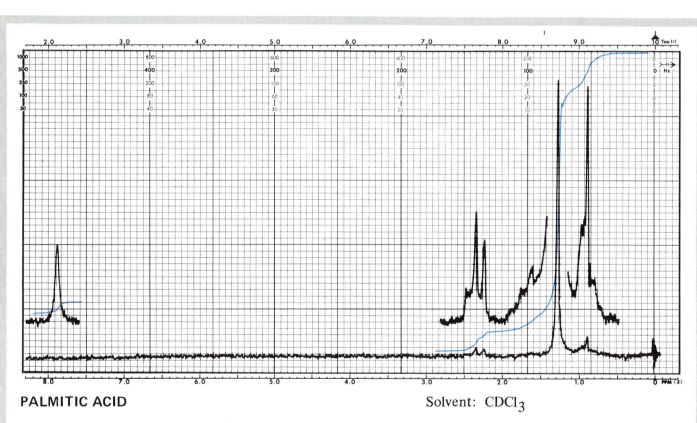

PALMITIC ACID

Solvent: $CDCl_3$

$$CH_3(CH_2)_{13}CH_2-\overset{\overset{\displaystyle O}{\|}}{C}-OH$$

Stearic acid which contains eighteen carbons is the longest of the commonly occurring saturated fatty acids. In comparison to the spectra of the three shorter acids (lauric, myristic, palmitic), stearic acid displays a more intense methylene band at 1.28 ppm and an integration value equivalent to thirty two hydrogens for the methylene absorbance bands in the range from 1.1 to 2.5 ppm.

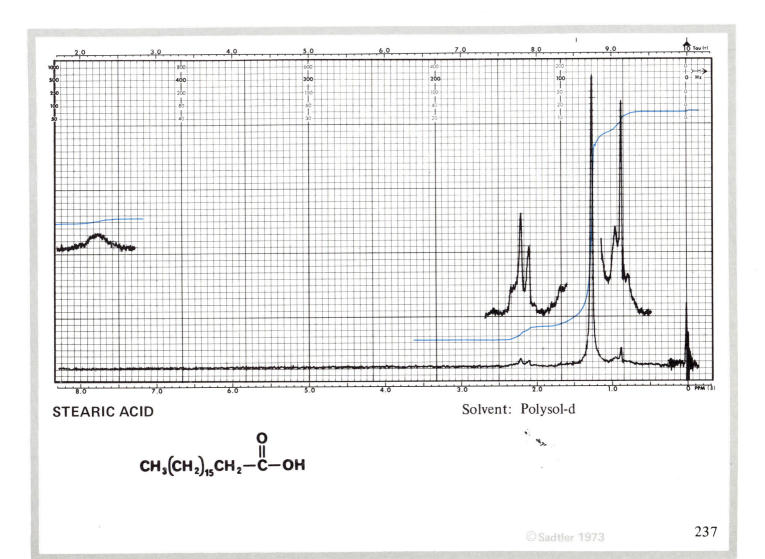

STEARIC ACID

Solvent: Polysol-d

$$CH_3(CH_2)_{15}CH_2-\overset{\overset{\displaystyle O}{\|}}{C}-OH$$

237

In addition to the bands at 0.89, 1.28 and 2.31 ppm, the unsaturated carboxylic acids display bands at 2.00 ppm arising from CH_2 groups adjacent to carbon-carbon double bonds and at 5.32 ppm which represents the resonance of the olefinic protons in the chain. The weak band at 2.77 ppm represents CH_2 groups bonded to two unsaturated carbons ($C=C-CH_2-C=C$) and indicates the presence of linoleic acid as an impurity in this sample. The most characteristic band of the unsaturated fatty acid chains is the weak triplet at about 5.3 ppm. The carboxylic acid proton appears on the offset as a sharp singlet and has a chemical shift of 11.5 ppm.

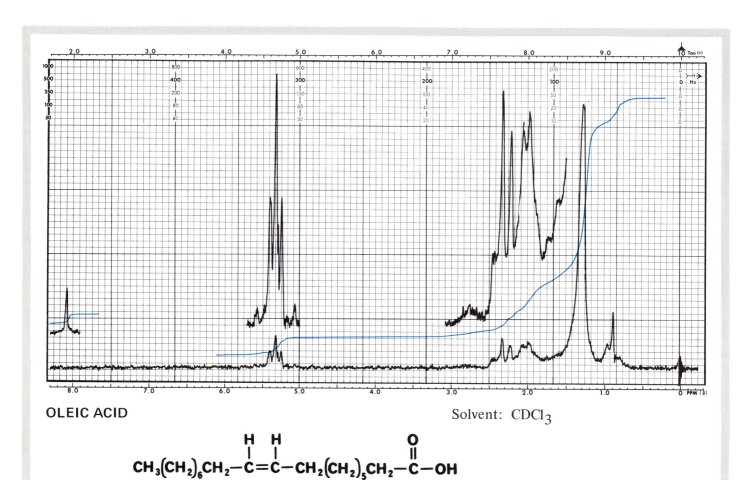

OLEIC ACID   Solvent: $CDCl_3$

238

Dimer acid is formed by the reaction of two unsaturated fatty acid molecules which are joined by connecting bonds linking the unsaturated carbons of the two acid chains. When the olefinic protons are converted to aliphatic hydrogens during the reaction, the characteristic olefinic absorbance bands at low field are lost and the resulting dimer acid spectrum becomes very similar in appearance to that of a simple saturated fatty acid.

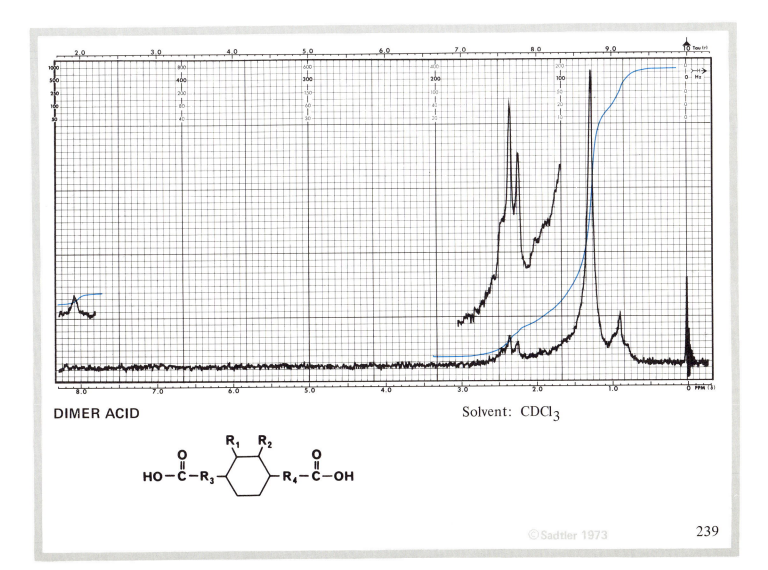

DIMER ACID

Solvent: $CDCl_3$

239

The molecular structure of trimer acid is analogous to that of dimer acid. As the name implies, trimer acid is produced by the condensation of three unsaturated fatty acid molecules to produce a single complex molecule containing three carboxylic acid groups. The appearance of the NMR spectrum is similar to that of dimer acid and those of simple saturated fatty acids. All of the bands, however, display considerable broadening due to the high molecular weight and complexity of the resulting trimer acid structure.

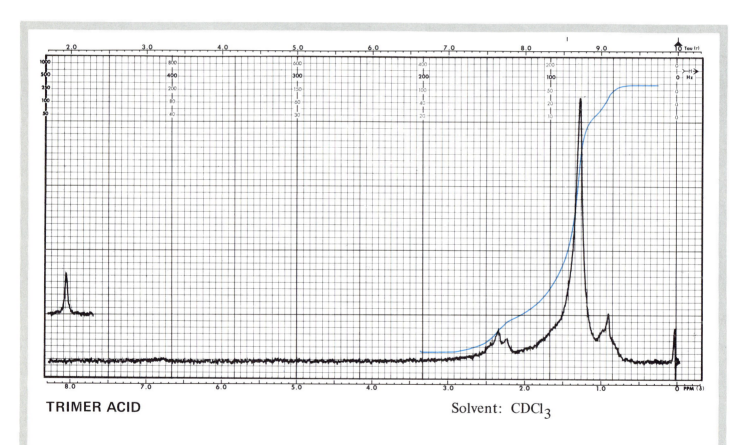

TRIMER ACID

Solvent: CDCl$_3$

COMPLEX STRUCTURE

The proton absorbance bands in this spectrum represent the major component of castor oil, glycerol tri-ricinoleate. In comparison with the spectrum of oleic acid, additional multiplets are observed at 2.59, 3.58, 4.23 and 5.22 ppm. The band at 2.58 ppm represents the hydroxyl group situated on the chain, the band at 3.58 ppm arises from the methine bonded to the hydroxyl oxygen atom, the band at 4.23 ppm represents the two glycerol methylene groups and the weak multiplet centered at about 5.22 ppm arises from the resonance of the glycerol methine protons.

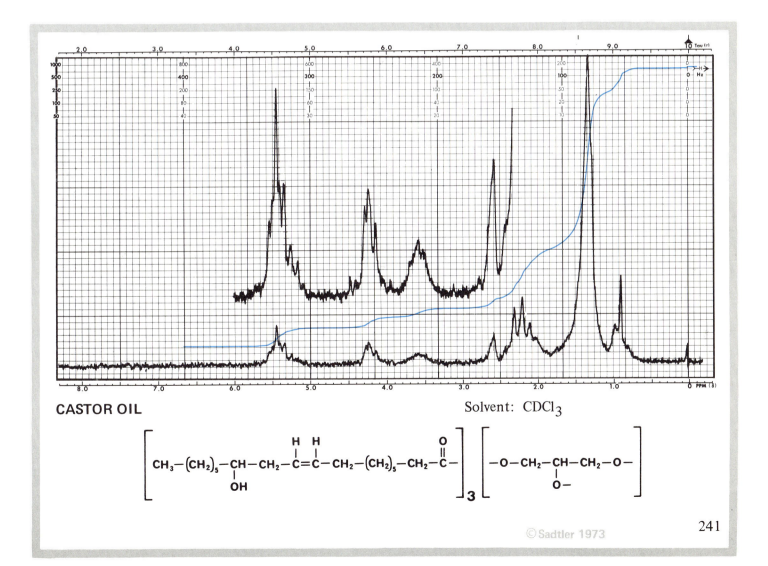

CASTOR OIL

Solvent: CDCl$_3$

241

Abietic acid is a major component of rosin and tall oil and is used in the preparation of many resins and plasticizers. The isopropyl methyl groups resonate at highest field as a doublet centered at about 0.89 ppm. The methyl groups bonded directly to the phenanthrene ring system appear as sharp singlets at 1.08 and 1.28 ppm. Numerous impurities are present in this commercial material as evidenced by the many bands in the olefinic/aromatic proton range from 4.6 to 7.5 ppm. The carboxylic acid protons present in the sample appear as a broad band with a chemical shift of about 10.38 ppm.

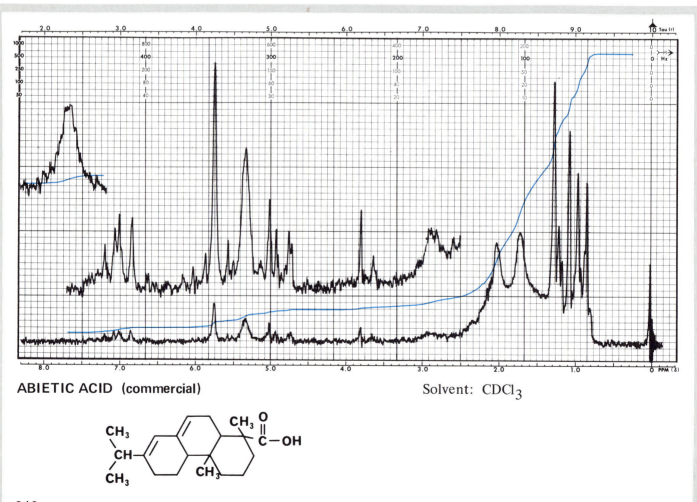

ABIETIC ACID (commercial)

Solvent: CDCl$_3$

Alpha Pinene is a monomer for the polyterpene resins which are utilized in the formulation of paints, varnishes, inks and adhesives. The strongest bands in the spectrum arise from the three non-equivalent methyl groups. The two geminal methyl groups which are non-equivalent resonate at 0.88 and 1.29 ppm as sharp singlets. The methyl group attached to cyclic double bond resonates at 1.65 ppm as a narrow multiplet. The methylene and methine groups in the ring system form a complex overlapping band in the chemical shift range from 1.8 to 2.6 ppm. The olefinic hydrogen appears at lowest field as a narrow multiplet displaying long range coupling to the methyl group and other nearby protons.

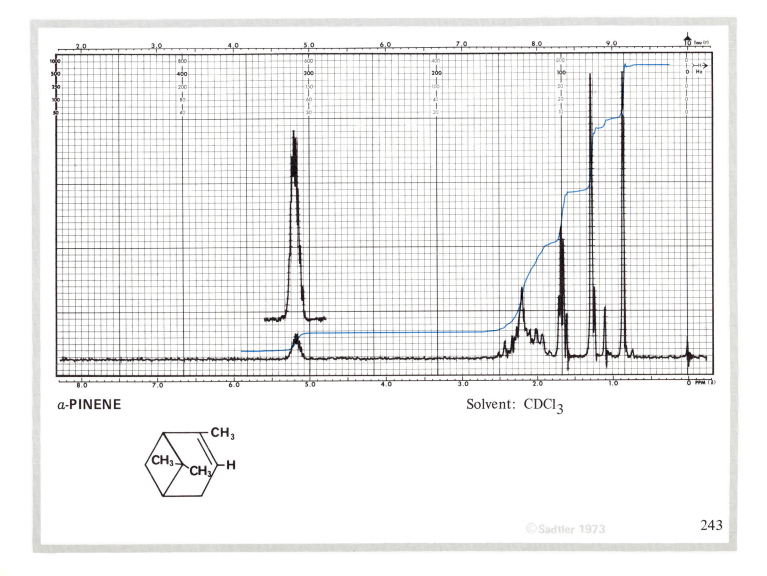

a-PINENE

Solvent: CDCl$_3$

243

The spectrogram of beta pinene resembles that of the alpha isomer only in the appearance of two intense sharp methyl singlets at high field. The chemical shift of these two peaks, 0.75 and 1.26 ppm are slightly different from the values observed for the alpha isomer. The ring methylenes and methines produce a complex series of peaks in the chemical shift range from 1.3 to 2.7 ppm. The olefinic hydrogen appears as a narrow multiplet centered at about 4.58 ppm. Both isomers are obtained from the distillation of turpentine with the beta isomer obtained at higher yields.

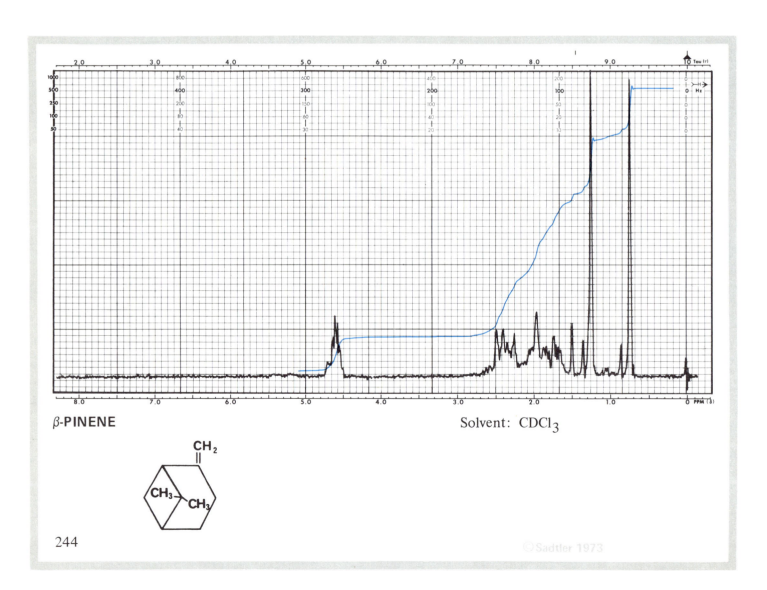

β-PINENE

Solvent: CDCl₃

244

This commercial sample contains a high proportion of dipentene (limonene) in addition to other terpenes and similar compounds. Dipentene is of interest since alpha-pinene undergoes a conversion to dipentene during the formation of alpha-pinene resins. The absorbance bands arising from dipentene appear as a doublet-like multiplet at 1.68 ppm, a broad band at 1.98 ppm and a narrow single peak at 4.67 ppm. The other bands are due to related terpene compounds with the exception of the peaks at 2.28 and 6.99 ppm which are due to the presence of xylene in the sample.

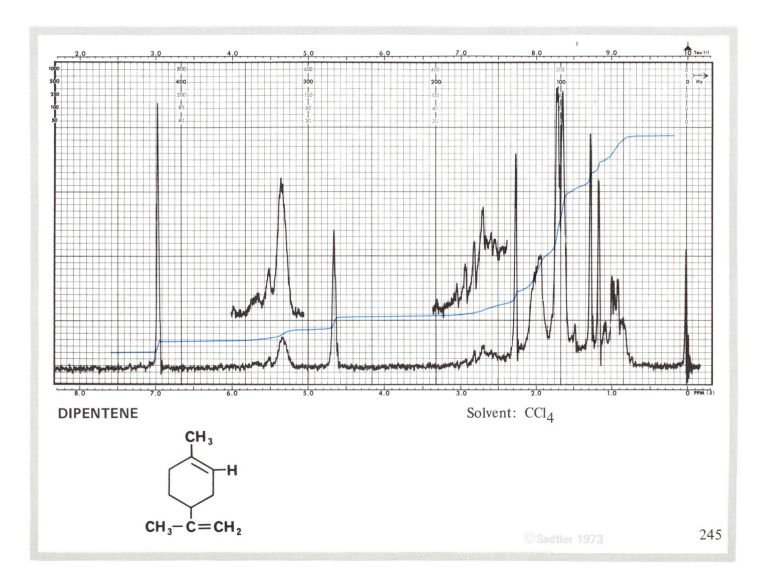

DIPENTENE

Solvent: CCl$_4$

245

Tall oil consists of about 60% of fatty acids and about 40% of rosin acids, of which abietic acid is a major component. The fatty acid content of this sample is indicated by the major peak at 1.28 ppm representing the CH_2 groups in the center of the fatty acid chain. The bands at 2.78 and 5.34 ppm indicate the presence in the chain of unsaturation of the linoleic type ($R-CH_2-CH=CH-CH_2-CH=CH-CH_2-R_1$).

The presence of rosin acids in the sample is indicated by the peaks at 0.86, 0.89, 1.09, 1.18 and 1.22 ppm which represent the methyl and isopropyl groups bonded to the rosin acid phenanthrene ring system. The "polymerization" of this sample was probably accomplished by sulfuric acid dehydration which converts Tall oil into a drying oil.

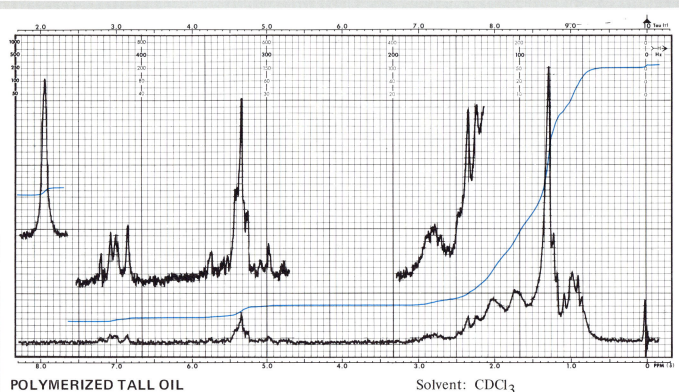

POLYMERIZED TALL OIL　　　　　　　　　　　　　Solvent: $CDCl_3$

COMPLEX STRUCTURE

This spectrum of commercial rosin acid is similar in composition to that of tall oil in that both contain fatty acids and rosin acids. The rosin acids possess a phenanthrene nucleus and have the general formula $C_{19}H_{29}COOH$. The presence of fatty acids in this spectrum is indicated by the bands at 1.29, 2.83 and 5.34 ppm. The rosin acid content is indicated by the sharp peaks at 0.85, 1.09, 1.18 and 1.21 ppm (compare with abietic acid on page 242). The aromatic pattern in the chemical shift range from 6.8 to 7.3 ppm arises from the presence of phenanthrene nuclei with an aromatic "A" ring.

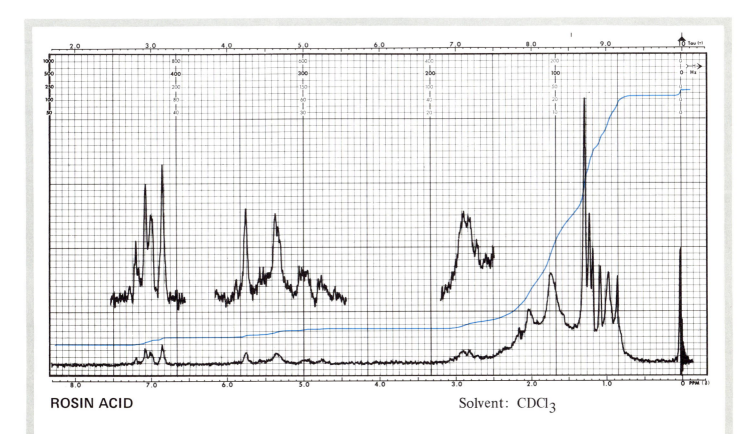

ROSIN ACID

Solvent: $CDCl_3$

COMPLEX STRUCTURE

247

The methacrylate protons resonate as follows; the methyl groups appear at 1.95 ppm as a broadened singlet, the olefinic proton which is trans to the carbonyl group resonates at 5.56 and the olefinic proton which is cis to the carbonyl resonates at 6.11 ppm.

The 1,3-butanediol methyl group resonates as a sharp doublet at 1.32 ppm. The center methylene group appears as a quartet centered at 2.0 ppm. The CH_2-O group resonates at 4.23 ppm as a triplet. The CH–O group appears at 5.13 ppm as a sextet.

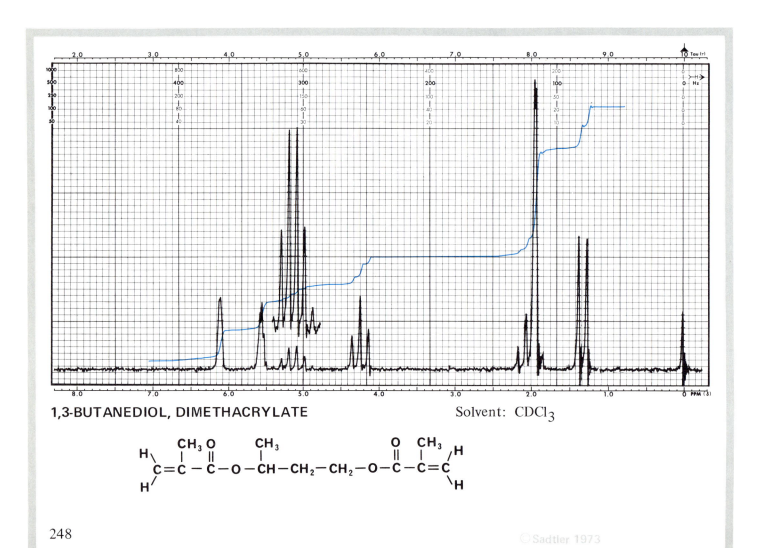

1,3-BUTANEDIOL, DIMETHACRYLATE   Solvent: $CDCl_3$

Diallyl phthalate is used as a monomer in the formation of allyl resins and also as a cross-linking agent in other polymer systems. The four aromatic hydrogens form a symmetrical AA'BB' pattern at lowest field. The hydrogens ortho to the carbonyl substituents resonate at slightly lower field than those which are meta. The olefinic methine groups appear at 6.0 ppm. The two terminal olefinic protons overlap in the range from 5.1 to 5.6 ppm. The allyl methylene groups appear at 4.8 ppm as a doublet, displaying additional long range coupling to the terminal vinyl protons.

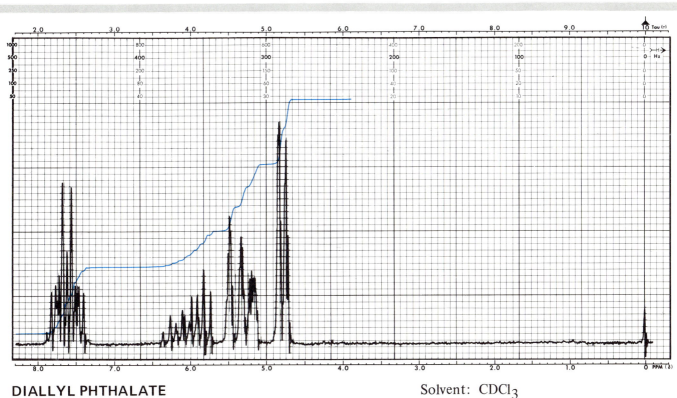

DIALLYL PHTHALATE

Solvent: CDCl$_3$

249

The allyl resonance in this spectrum is virtually identical to that of diallyl phthalate which is depicted on the opposite page. The appearance of the aromatic patterns is quite different however, with the phthalate producing a complex symmetrical pattern while the terephthalate ring hydrogens, which are equivalent, resonate as a single sharp peak at 8.12 ppm.

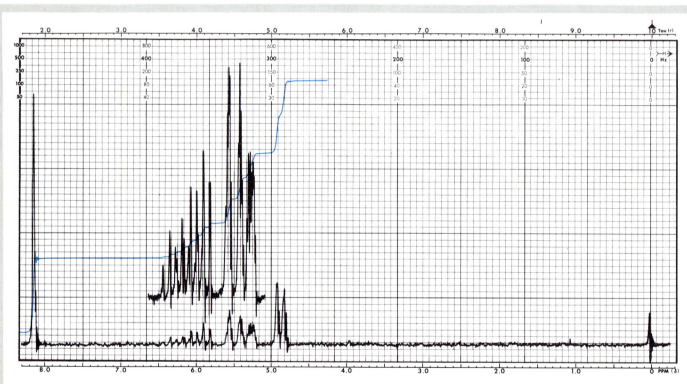

DIALLYL TEREPHTHALATE

Solvent: $CDCl_3$

$$H_2C=CH-CH_2-O-\overset{\overset{\displaystyle O}{\|}}{C}--\overset{\overset{\displaystyle O}{\|}}{C}-O-CH_2-CH=CH_2$$

250

1,4-Cyclohexanedimethanol is used as a monomer to form polyesters with terephthalic acid (page 109) which are used as textile fibers. This spectrogram depicts the di(trifluoroacetate) derivative since the appearance of its high field pattern more closely matches that of the monomer as it appears in polymer spectra than that of the diol. The ten cyclohexane ring hydrogens resonate as a complex envelope of absorbance in the chemical shift range from 0.9 to 2.3 ppm. The two methylene groups adjacent to the ring appear at 4.29 ppm as a clear doublet. The doublet multiplicity of this band indicates that the two methanol substituents are situated trans to each other on the ring.

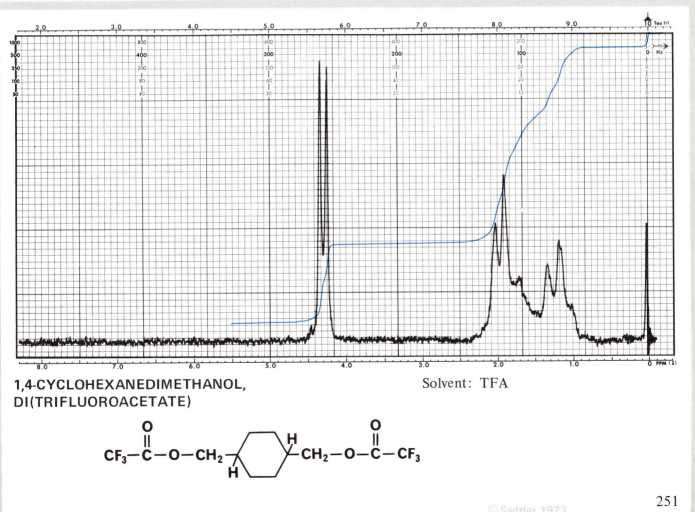

1,4-CYCLOHEXANEDIMETHANOL, DI(TRIFLUOROACETATE)

Solvent: TFA

251

The four centrally located methylene groups resonate at highest field forming a characteristically broad band at about 1.4 ppm. The two terminal methylene groups which are bonded to the nitrogen atoms occur as a broadened triplet at 2.7 ppm. When incorporated into a polymer such as nylon 6,6, the methylene **groups** bonded to nitrogen resonate at lower field near 3.5 ppm instead of 2.7 ppm as they do in the monomer.

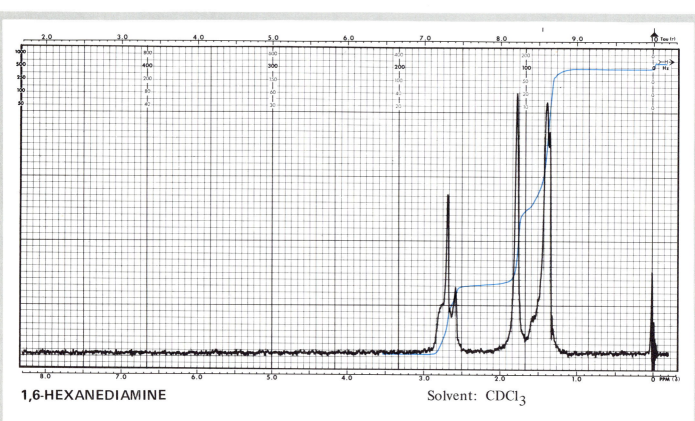

1,6-HEXANEDIAMINE　　　　　　　　　　Solvent: $CDCl_3$

$$NH_2-CH_2(CH_2)_4CH_2-NH_2$$

Although not magnetically nor chemically equivalent, the four aromatic protons possess very similar chemical shifts and they resonate as a single sharp peak. The four amine protons are equivalent and resonate as a singlet at 3.3 ppm.

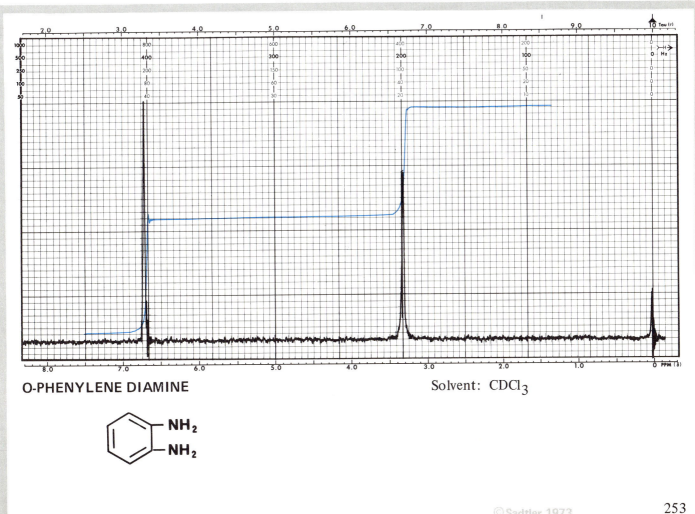

O-PHENYLENE DIAMINE

Solvent: CDCl₃

253

The three protons which are ortho to one or more of the amine nitrogen substituents overlap to form a complex multiplet near 6.05 ppm (H-2, H-4, H-6). The ring hydrogen which is meta to the two nitrogen atoms (H-5) appears at lower field as a diortho triplet near 6.93 ppm. The four amine protons resonate at 3.48 ppm as a broad band.

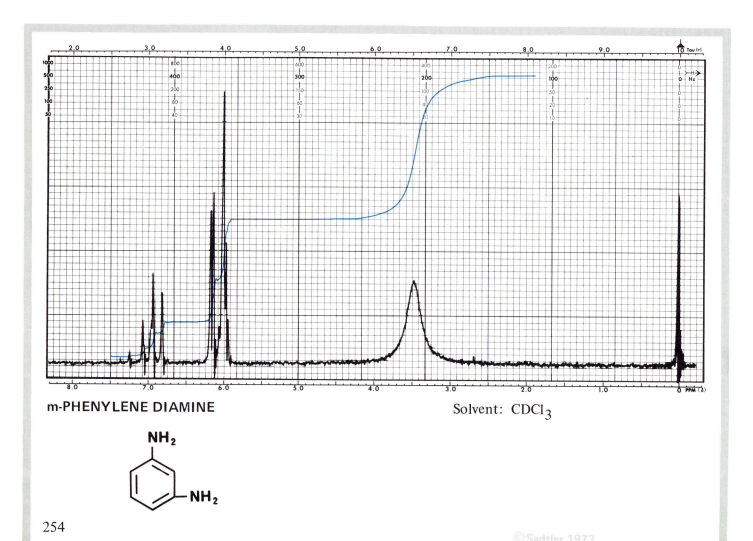

m-PHENYLENE DIAMINE

Solvent: CDCl₃

The four equivalent aromatic protons of p-phenylene diamine resonate at 6.58 ppm as a sharp singlet. The four amine protons absorb at 3.21 ppm as a broad singlet.

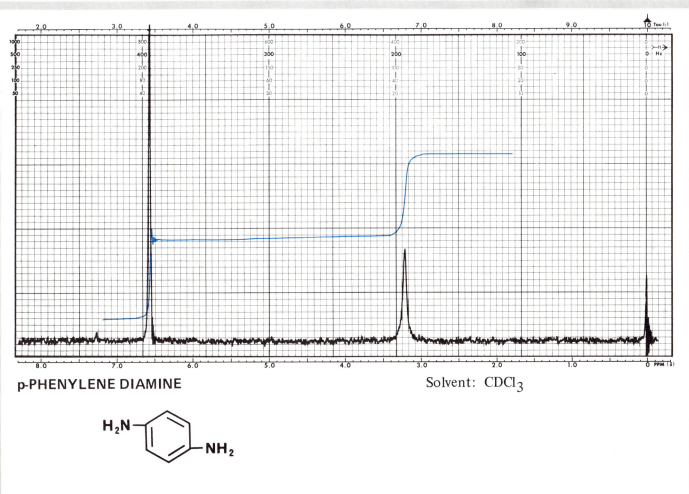

p-PHENYLENE DIAMINE

Solvent: $CDCl_3$

255

Benzoguanamine (2-phenyl melamine) is the amine used in the preparation of the amine-formaldehyde resin presented on page 156. The NMR spectrum of the monomer or that of the resin, provide no direct indication for the presence of the melamine ring system. In this spectrum, the broad band which resonates at 6.62 ppm was found to be exchangeable and represents the two primary amine groups. The band centered at 7.38 ppm is attributable to the para and meta protons of the phenyl substituent while the band at lower field (8.25 ppm), represents the two protons ortho to the nitrogen substituent. The secondary amine group which joins the melamine and benzene ring systems, was not observed. Exchangeable protons of this type often appear as a very broad, weak band at low field.

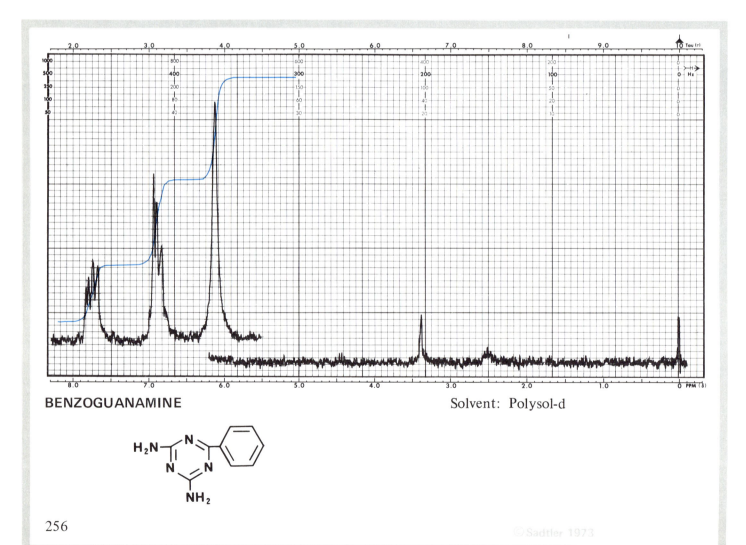

BENZOGUANAMINE

Solvent: Polysol-d

The four methylene groups of adipamide resonate as two broadened multiplets with chemical shifts of 1.90 and 2.67 ppm. The multiplet at higher field represents the two equivalent methylene groups in the center of the molecule while the band at lower field is attributable to the CH_2 groups which are bonded to the amide carbonyl groups. The poor resolution which is observed for these two bands is usually apparent when trifluoroacetic acid is used as a solvent (compare with the spectrum of adipic acid on page 229). The four amide protons resonate below 8.3 ppm and overlap with the carboxylic acid protons of the solvent.

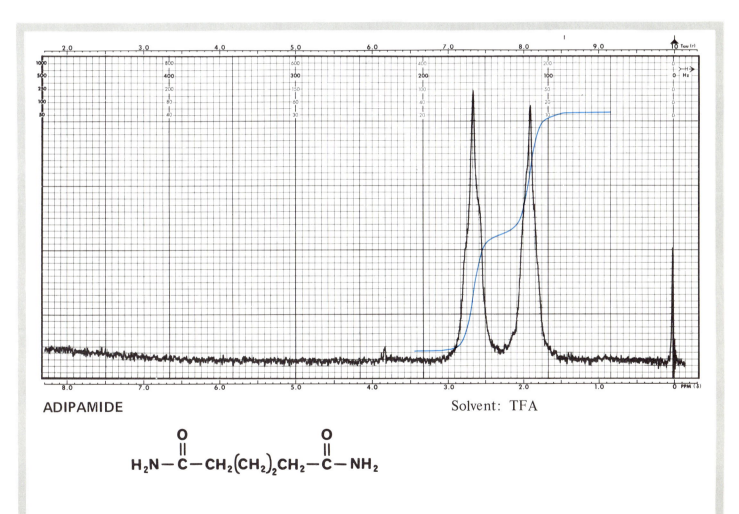

ADIPAMIDE   Solvent: TFA

$$H_2N-\overset{\overset{\displaystyle O}{\|}}{C}-CH_2(CH_2)_2CH_2-\overset{\overset{\displaystyle O}{\|}}{C}-NH_2$$

257

MONOMERS

In this spectrum of the nylon 6 monomer, the amide proton appears at lowest field as a very broad band centered at 7.3 ppm. The methylene groups which are bonded to the nitrogen atoms and carbonyl groups resonate at 3.15 ppm and 2.38 ppm respectively. The remaining ring methylene groups form a broad band centered at 1.7 ppm. Upon polymerization, these proton groups undergo a shift to lower field (see page 118) due in part to the solvents used to form the polymer solutions; trifluoroacetic acid and formic acid.

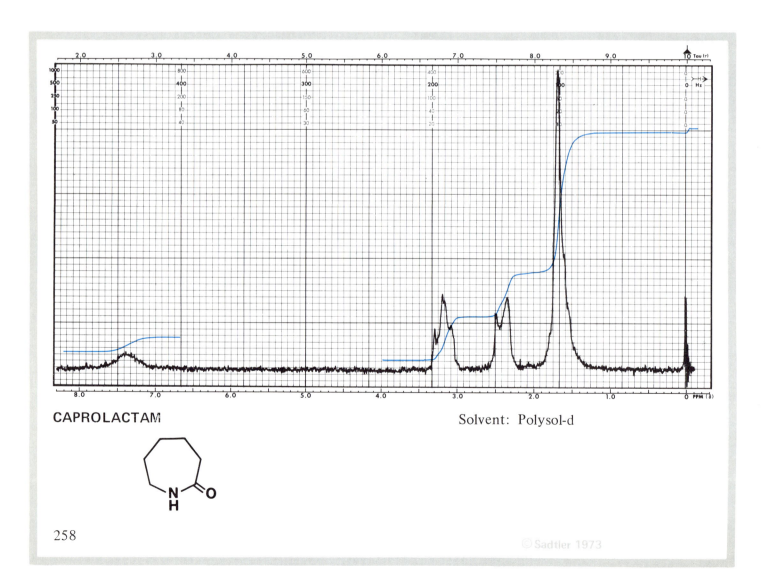

CAPROLACTAM

Solvent: Polysol-d

258

The four exchangeable protons of urea resonate as a broad band near 5.4 ppm. Upon substitution of one of the nitrogen atoms by an alkyl group, the remaining amide hydrogen undergoes a shift to lower field and resonates in the 6 to 8 ppm range. Bands of this type are observed in the NMR spectra of polyamides and polyurethanes (p. 126). Amide protons are exchangeable and will display a gradual shift to higher field with increasing sample solution temperature.

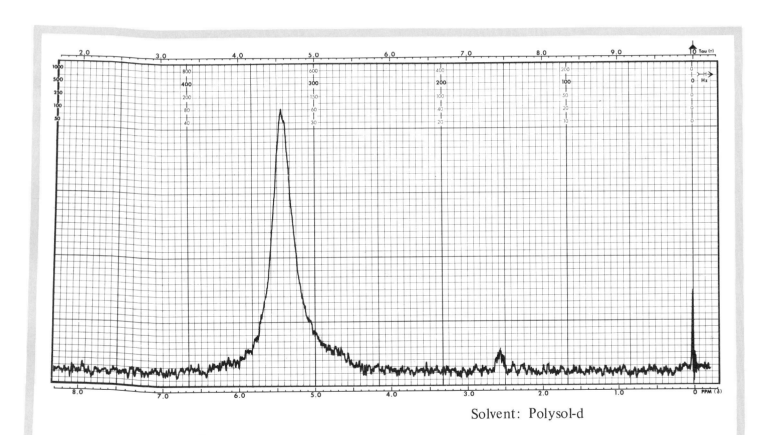

Solvent: Polysol-d

259

As a monomer for the sulfonamide-formaldehyde resin on page 157, para toluene sulfonamide displays a relatively uncomplicated spectrogram consisting of two single peaks and a low-field para substituted benzene pattern. The three hydrogen singlet at 2.37 ppm represents the toluene methyl group. The broad two proton band at 6.98 ppm arises from the two exchangeable sulfonamide protons. The four aromatic protons appear as the two ortho doublets. The doublet at 7.23 ppm is broadened by long range coupling to the adjacent methyl group. The protons ortho to the sulfonyl group appear at lower field with a chemical shift of 7.72 ppm.

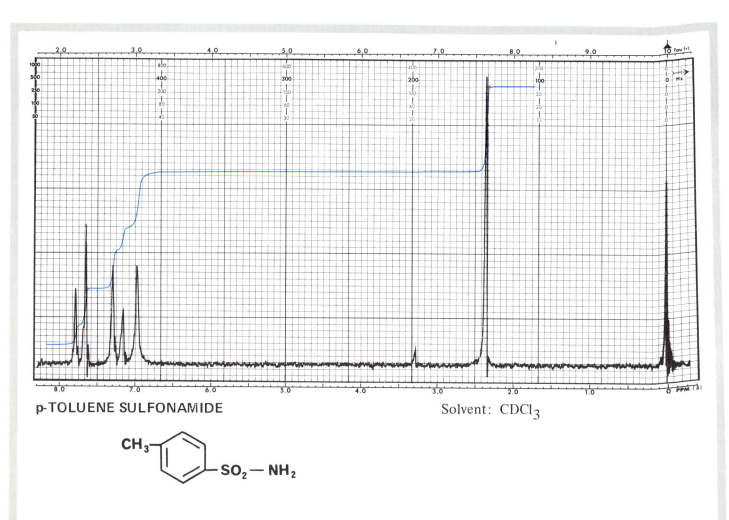

p-TOLUENE SULFONAMIDE

Solvent: $CDCl_3$

CH_3 — benzene ring — $SO_2 - NH_2$

This cyclic tetramer of dimethyl siloxane is used to prepare high molecular weight linear polydimethyl siloxane polymers. The eight methyl groups, if not actually equivalent, possess such similar chemical shifts that they resonate as a sharp singlet just downfield from TMS at 0.10 ppm.

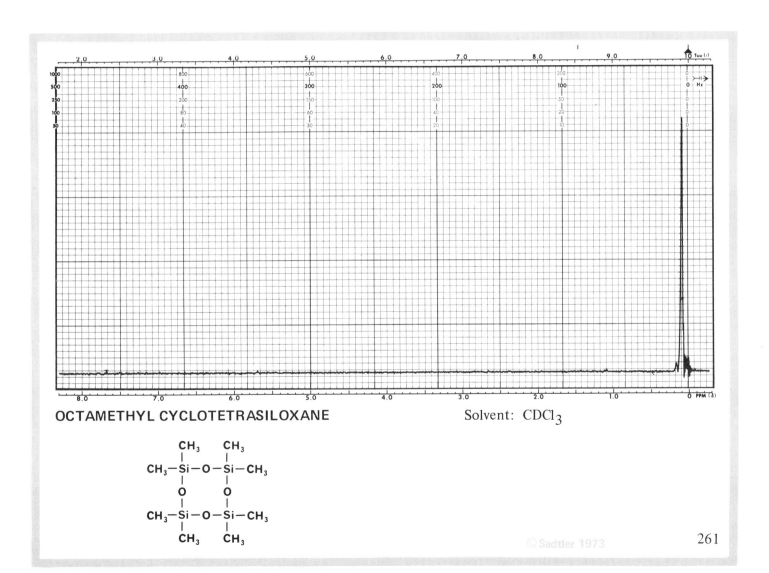

OCTAMETHYL CYCLOTETRASILOXANE   Solvent: $CDCl_3$

© Sadtler 1973

Dichloro diphenyl silane is a monomer for certain of the polysiloxanes which contain phenyl groups (see page 167, 168). Although the effect of silicone on aliphatic groups is that of a strongly shielding substituent, its effect upon aromatic protons is that of a moderately strong <u>deshielding</u> group. The two hydrogens which are ortho to the silicone atom resonate at lowest field near 7.73 ppm. The para and meta hydrogens overlap to form a complex multiplet in the chemical shift range from 7.1 to 7.55 ppm.

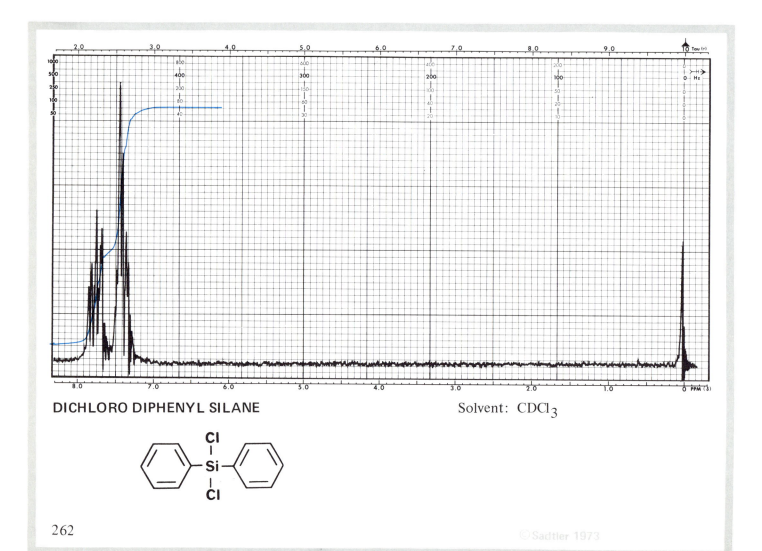

DICHLORO DIPHENYL SILANE

Solvent: $CDCl_3$

The polymer finder chart has been designed to aid the spectroscopist in the identification and analysis of polymer NMR spectra. Every effort has been made to make the chart of general applicability, although individual samples from other sources may not correspond exactly to those used to develop the table. It is hoped that through use, and with the suggestions and comments of the users, that the scope and accuracy of this chart may be expanded and improved.

Chart Headings

1. Chem. Shift: This value corresponds to the chemical shift in ppm (δ - scale) for one band of the polymer. Each major band in the spectrum is listed in order of decreasing field (increasing chemical shift).

2. Band Type: This column contains a descriptive definition of the appearance of the band in the spectrum. The following list indicates the abbreviations that are used:

S	=	singlet	Q	=	quartet
SS	=	sharp singlet	M	=	multiplet
BS	=	broad singlet	CM	=	complex multiplet
D	=	doublet	BM	=	broad multiplet
T	=	triplet	BB	=	broad band

3. Structure: The structures in this column depict in **heavy type** the proton groups and the environment which produces the band being described.

4. Other Bands: In this section are listed the other major bands in the spectrum belonging to the monomer being described. For example, with monomers which occur in several different types of polymers (e.g. adipic acid in esters, amides, urethanes) only the adipate bands are listed, the alcohol or amine fragments would have to be traced independently. Bands which overlap in the spectrum are indicated by underlining their chemical shift values, i.e., 1.35 1.76 5.23.

5. Polymer Type: Here is presented either the name of the specific polymer or the polymer fragment being described.

6. Solvent: This column lists the solvent in which the spectrum was run. It should be compared to that used by the spectroscopist since solvent shifts can be appreciable. In a few instances, the table contains listings for the same polymer in two different solvents.

263

7. <u>Page:</u>   Refers to the page on which a typical example of the polymer or fragment pattern is presented.

How to use the polymer finder chart

1. Select the most characteristic band from the unknown spectrum. These bands are often low field peaks (e.g. olefinic or aromatic bands) since many polymers exhibit high field (saturated aliphatic) bands.

2. Find the chemical shift of this band in the first column allowing for a deviation of ± 0.3 ppm for solvent and minor compositional differences.

3. Select those listings which have the same appearance (<u>Band type</u>) as the unknown band.

4. Compare the remaining bands in the spectrum with those listed under <u>Other Bands.</u>

5. Select listing with the same or a similar solvent to that of the unknown spectrum. Compare unknown spectrum with example in text.

6. Any bands which remain after these steps are probably due to either a co-monomer, copolymer, plasticizer, solvent or an impurity. Co-monomers and copolymers should be separately analyzed as above.

264

CHEM SHIFT	BAND TYPE	STRUCTURE	OTHER BANDS	POLYMER TYPE	SOLVENT	PAGE
0.09	SS	$(CH_3)_2-Si-O-$	- - - -	DIMETHYL SILOXANE	$CDCl_3$	161
0.13	SS	CH_3-Si-O	0.18 0.78 2.02	FLUOROSILICONE	$CDCl_3$	164
0.18	SS	CH_3-Si-O	0.13 0.78 2.02	FLUOROSILICONE	$CDCl_3$	164
0.19	SS	CH_3-Si-O	0.45	SILICONE RUBBER	TFA	162
0.70	S	$(CH_3)_3-C$	1.3 1.69 7.0	OCTYLPHENOL	$CDCl_3$	154
0.84	D	$(CH_3)_2-CH-$	1.1 1.5	POLYPROPYLENE	$CDCl_3$	54
0.88	BM	CH_3-CH_2-	1.0 1.32	POLYBUTENE	CCl_4	32
0.90	BS	CH_3-C	1.05 1.85 3.6	POLYMETHACRYLATE	$CDCl_3$	64
0.90	BS	CH_3-C	1.9 4.07	POLYACRYLATE	$CDCl_3$	62
0.90	BS	CH_3-C	1.27 1.86 4.0	POLYMETHACRYLATE	$CDCl_3$	66
0.90	D	$(CH_3)_2-CH-$	1.1 1.6	POLYMETHYL PENTENE	$CDCl_3$	30
0.91	D	$(CH_3)_2-CH-CH_2-O$	1.7 3.2 3.5	POLYVINYL ETHER	$CDCl_3$	60
0.92	D	$(CH_3)_2-CH-CH_2-O$	1.95 3.75	POLYMETHACRYLATE	$CDCl_3$	68
0.92	T	CH_3-CH_2-CH-O	1.5 3.35	1,2-BUTYLENE GLYCOL	CCl_4	93
0.95	BM	CH_3-C	1.5 1.85 3.96	POLYMETHACRYLATE	CCl_4	67
0.95	T	CH_3-CH_2	1.6 3.95 4.6	POLYVINYL BUTYRAL	$CDCl_3$	80
0.98	T	$CH_3-CH_2CH_2CH_2-O-$	1.5 2.18 3.96	BUTYL POLYACRYLATE	CCl_4	63
1.00	M	$-CH_2CH_2-$	0.88 1.32	POLYBUTENE	CCl_4	32
1.05	S	CH_3-C	0.90 1.85 3.6	POLYMETHACRYLATE	$CDCl_3$	64
1.10	M	$-CH_2CH_2-$	0.90 1.60	POLYMETHYL PENTENE	$CDCl_3$	30
1.10	M	$-CH_2-CH-CH_3$	0.84 1.50	POLYPROPYLENE	$CDCl_3$	28
1.10	SS	$(CH_3)_2-C$	1.40	POLYISOBUTYLENE	CCl_4	29
1.13	D	CH_3-CH	3.42	PROPYLENE GLYCOL	CCl_4	90
1.19	T	CH_3-CH_2-O-	1.65 3.48	POLYVINYL ETHER	$CDCl_3$	59

CHEM SHIFT	BAND TYPE	STRUCTURE	OTHER BANDS	POLYMER TYPE	SOLVENT	PAGE
1.22	SS	$(CH_3)_3-\emptyset$	7.01	t-BUTYLPHENOL	$CDCl_3$	152
1.25	D	$CH_3-CH-O-$	1.84 4.08 4.95	1,3-BUTYLENE GLYCOL	CCl_4	105
1.25	BB	$C-CH_2-C$	1.78 4.76	BUTADIENE	$CDCl_3$	37
1.27	T	$CH_3-CH_2-O-C(=O)-$	0.90 1.86 4.00	METHACRYLATE	$CDCl_3$	66
1.27	SS	$-CH_2-CH_2-CH_2-$	0.90 (weak)	POLYETHYLENE	$CDCl_3$	26
1.30	BB	$-CH_2(CH_2)_n-CH_2-$	1.50 2.63 3.50	NYLON 6, 10	FA	120
1.31	S	$(CH_3)_2-C$	0.70 1.69 7.00	OCTYLPHENOL	$CDCl_3$	154
1.32	M	$-CH-$	0.88 1.00	POLYBUTENE	CCl_4	32
1.33	SS	$-CH_2-$	- - - -	POLYETHYLENE	ODCB	27
1.35	SS	Cyclohexyl	2.00 4.45	POLYESTER	TFA	109
1.40	BS	$-CH_2(CH_2)_n-CH_2-$	1.70 2.75 3.60	NYLON 11 or 12	TFA	122
1.41	SS	$C-CH_2-C$	1.10	ISOBUTYLENE	CCl_4	29
1.45	M	$-CH_2-CH_2-CH_2-$	1.70 2.65 3.50	NYLON 6, 10	TFA	- - -
1.50	BM	$-CH-$	0.84 1.10	POLYPROPYLENE	$CDCl_3$	28
1.50	BB	$-CH_2-CH-\emptyset$	1.85 6.60 7.06	POLYSTYRENE	$CDCl_3$	42
1.50	BB	$-CH_2-CH-\emptyset$	2.00 6.60 7.06	POLYSTYRENE	$CDCl_3$	43
1.50	BM	$CH_3-CH_2CH_2-CH_2-O$	0.98 2.18 3.96	BUTYL POLYACRYLATE	CCl_4	63
1.50	Q	$-CH_2-CH_2-CH-O-$	0.92 3.35	1,3-BUTYLENE GLYCOL	CCl_4	93
1.50	BB	$-CH_2-(CH_2)_4-CH_2-$	1.85 2.70 3.50	NYLON 6, 6	TFA	119
1.50	BB	$-CH_2(CH_2)_4CH_2-$	1.30 2.60 3.50	NYLON 6, 10	FA	120
1.55	BB	$CH_3-CH_2CH_2-CH_2-O-$	0.95 1.85 3.96	POLYMETHACRYLATE	$CDCl_3$	67
1.60	BB	$R-CH-R_2$	0.90 1.10	POLYMETHYL PENTENE	$CDCl_3$	30
1.60	BB	$-CH_2-CH_2-CH-O-$	0.95 4.00 4.60	POLYVINYL BUTYRAL	$CDCl_3$	80
1.60	BM	$-CH-CH_2-CH-$	3.75 4.85	POLYVINYL FORMAL	ODCB	78

CHEM SHIFT	BAND TYPE	STRUCTURE	OTHER BANDS	POLYMER TYPE	SOLVENT	PAGE
1.60	BM	$-CH-CH_2-CH-$	3.75 4.63	POLYVINYL FORMAL	$CDCl_3$	79
1.60	S	$(CH_3)_2-C-\emptyset_2$	7.0	BISPHENOL−A	$CDCl_3$	114
1.60	BB	$-CH_2-CH-\emptyset$	7.10	STYRENE COPOLYMER	$CDCl_3$	46
1.63	T	$O-CH_2(CH_2)_2CH_2-O$	3.4	POLYGLYCOL	$CDCl_3$	92
1.63	BS	$N-CH_2(CH_2)_4CH_2-N$	2.55	HDI URETHANE	$CDCl_3$	135
1.65	M	$-CH-CH_2-CH-O-$	1.17 3.48	VINYL ETHER	$CDCl_3$	59
1.68	T	$-CH_2(CH_2)_2CH_2-$	2.37	ADIPATE	$CDCl_3$	101
1.69	S	$C-CH_2-C$	0.70 1.31 7.00	OCTYLPHENOL	$CDCl_3$	154
1.70	BM	$-CH_2(CH_2)_3CH_2-N$	2.75 3.68	NYLON 6	TFA	118
1.70	BM	$-CH_2(CH_2)_nCH_2-$	1.45 2.65 3.50	NYLON 6, 10	FA	- - -
1.70	BM	$-CH-CH_2-CH-O$	0.91 3.20 3.50	VINYL ETHER	$CDCl_3$	60
1.70	Q	$-CH-CH_2-CH-O$	<u>3.35 3.45</u>	VINYL ETHER	$CDCl_3$	58
1.70	S	$(CH_3)_2-C-\emptyset_2$	7.00 7.80	BISPHENOL−A	$CDCl_3$	96
1.70	BB	$-CH-CH_2-CH-OH$	4.05	VINYL ALCOHOL	D_2O	56
1.70	BM	$-CH_2-CH_2-CH_2-NH-$	1.40 2.75 3.60	NYLON 11 or 12	TFA	122
1.73	CM	$-CH_2(CH_2)_2CH_2-O-$	2.78 3.52 4.62	POLYSULFIDE	$CDCl_3$	98
1.75	S	$CH_3-C=C$	2.12 5.25	POLYISOPRENE	$CDCl_3$	39
1.75	S	$CH_3-C=C$	2.12 5.25	NATURAL RUBBER	ODCB	40
1.75	T	$CH_2-(CH_2)_2CH_2$	2.55	ADIPATE	TFA	132
1.78	M	$-CH_2-CH-CH=CH_2$	1.25 <u>4.76 5.20</u>	1,2-BUTADIENE	$CDCl_3$	37
1.82	BB	$-CH-CH_2-CH-O-$	2.02 4.88	VINYL ACETATE	$CDCl_3$	61
1.82	BB	$-CH_2-CH-$Pyridine	6.35 6.79 7.12 7.69	VINYL PYRIDINE	$CDCl_3$	70
1.84	Q	$O-CH-CH_2-CH_2-O-$	1.25 4.08 4.95	1,3-BUTYLENE GLYCOL	CCl_4	105

CHEM SHIFT	BAND TYPE	STRUCTURE	OTHER BANDS	POLYMER TYPE	SOLVENT	PAGE
1.85	BB	CH$_2$—CH—Ø	1.50 6.60 7.06	POLYSTYRENE	CDCl$_3$	42
1.85	BM	—CH$_2$—**CH$_2$**—CH$_2$—N	1.50 2.70 3.50 8.30	NYLON 6, 6	TFA	119
1.85	BB	—C—**CH$_2$**—C—	0.90 1.05 3.60	METHACRYLATE	CDCl$_3$	64
1.85	BB	C—**CH$_2$**—C	0.90 1.27 4.00	METHACRYLATE	CDCl$_3$	66
1.85	BB	C—**CH$_2$**—C	0.95 1.55 3.96	METHACRYLATE	CDCl$_3$	67
1.87	BS	CH$_2$—**CH$_2$**—CH$_2$	4.3	POLYURETHANE	TFA	133
1.90	BB	C—**CH$_2$**—C	0.90 1.06 3.60	METHACRYLATE	CDCl$_3$	64
1.95	BB	C—**CH$_2$**—C	0.92 3.75	METHACRYLATE	CDCl$_3$	68
2.00	BB	—CH$_2$—**CH**—Ø	1.50 6.60 7.06	POLYSTYRENE	CDCl$_3$	43
2.00	CM	Cyclohexyl	1.35 4.45	POLYESTER	TFA	109
2.02	CM	Si—CH$_2$**CH$_2$**—CF$_3$	0.13 0.18 0.88	FLUOROSILICONE	CDCl$_3$	164
2.02	SS	**CH$_3$**—C(=O)—O	1.82 4.88	POLYVINYL ACETATE	CDCl$_3$	61
2.09	BS	CH=CH—**CH$_2$**—CH=CH	5.40	POLYBUTADIENE	CDCl$_3$	35
2.10	BB	—CH—**CH$_2$**—CH—**CN**	3.15	POLYACRYLONITRILE	DMSO-d6	52
2.11	S	**CH$_3$**—Ø—	6.50	POLYPHENYLENE OXIDE	CDCl$_3$	95
2.12	CM	C—**CH$_2$**—C=CH—	1.75 5.25	POLYISOPRENE	CDCl$_3$	39
2.12	CM	C—**CH$_2$**—C=CH—	1.75 5.25	NATURAL RUBBER	ODCB	40
2.14	BB	CH—**CH$_2$**—CH—Cl	4.48	VINYL CHLORIDE	ODCB	54
2.18	BB	—CH$_2$—**CH**—C(=O)—O	0.98 1.50 3.96	ACRYLATE	CCl$_4$	63
2.25	BB	Pyrrolidone	3.30 3.95	VINYL PYRROLIDONE	CDCl$_3$	69
2.37	T	CH$_2$—**CH$_2$**—C(=O)—O	1.68	ADIPATE	CDCL$_3$	101
2.53	T	CH$_2$—**CH$_2$**—C(=O)—O	1.75	ADIPATE	TFA	132
2.60	T	CH$_2$—**CH$_2$**—C(=O)—NH—	<u>1.30 1.50</u> 3.50	NYLON 6, 10	FA	120

CHEM SHIFT	BAND TYPE	STRUCTURE	OTHER BANDS	POLYMER TYPE	SOLVENT	PAGE
2.65	CM	Epoxide	3.30 4.00	GLYCIDYL ETHER	$CDCl_3$	142
2.65	M	Epoxide	3.15 3.60	GLYCIDYL ETHER	$CDCl_3$	139
2.65	BM	CH_2–**CH_2**–C(=O)–NH	<u>1.45 1.70</u> 3.50	NYLON 6, 10	FA	- - -
2.70	T	CH_2–**CH_2**–C(=O)–NH	<u>1.50 1.85</u> 3.50	NYLON 6, 6	TFA	119
2.75	BB	CH_2–**CH_2**–C(=O)–NH–	1.70 3.68	NYLON 6	TFA	118
2.75	T	CH_2–**CH_2**–C(=O)–NH	<u>1.40 1.75</u> 3.60	NYLON 11 or 12	TFA	122
2.78	T	CH_2–**CH_2**–S–	1.75 3.52 4.62	POLYSULFIDE	$CDCl_3$	98
2.95	T	–S–**CH_2**–CH_2–O–	3.91 4.75	POLYSULFIDE	$CDCl_3$	97
3.00	BM	–CCl_2–**CH_2**–CCl_2–	- - - -	VINYLIDENE CHLORIDE	$CDCl_3$	55
3.15	BB	–CH_2–**CH**–CN	2.10	ACRYLONITRILE	DMSO-d6	52
3.15	M	Epoxide	2.65 3.60	GLYCIDYL ETHER	$CDCl_3$	139
3.20	D	CH–**CH_2**–CH(CH_3)$_2$	0.91 1.70 3.50	POLYVINYL ETHER	$CDCl_3$	60
3.30	BB	Pyrrolidone	2.25 3.95	VINYL PYRROLIDONE	$CDCl_3$	69
3.30	CM	Epoxide	2.65 4.00	GLYCIDYL ETHER	$CDCl_3$	142
3.35	SS	**CH_3**–O–CH–CH_2	1.70 3.45	POLYVINYL ETHER	$CDCl_3$	58
3.35	CM	–O–**CH**(CH_3)–CH_2–**CH_2**–O–	0.92 1.50	1,3-BUTYLENE GLYCOL	CCl_4	93
3.40	T	(CH_2)$_2$–**CH_2**–O–	1.63	1,4-BUTYLENE GLYCOL	$CDCl_3$	92
3.45	BB	CH_2–**CH**–O–CH_3	1.70 3.35	POLYVINYL ETHER	$CDCl_3$	58
3.48	Q	CH_3–**CH_2**–O–CH–	1.17 1.65	POLYVINYL ETHER	$CDCl_3$	59
3.50	M	CH_2–O–**CH**(CH_2)$_2$	0.91 1.70 3.20	POLYVINYL ETHER	$CDCl_3$	60
3.50	M	(CH_2)$_n$–**CH_2**–NH–C(=O)–	1.30 1.50 2.60	NYLON 6, 10	FA	120
3.50	M	(CH_2)$_n$–**CH_2**–NH–C(=O)–	<u>1.45 1.7</u> 2.65	NYLON 6, 10	TFA	- - -
3.50	BB	(CH_2)$_n$–**CH_2**–NH–C(=O)–	<u>1.5 1.85</u> 2.7	NYLON 6, 6	TFA	119
3.52	T	–S–CH_2(CH_2)$_2$–**CH_2**–O–	1.73 2.78 4.62	POLYSULFIDE	$CDCl_3$	98

CHEM SHIFT	BAND TYPE	STRUCTURE	OTHER BANDS	POLYMER TYPE	SOLVENT	PAGE
3.58	BB	$(CH_3)_4$–CH_2–NH–C(=O)–	1.70 2.75	NYLON 6	TFA	118
3.60	BB	$(CH_2)_n$–CH_2–NH–C(=O)–	1.40 1.75 2.75	NYLON 11 or 12	TFA	122
3.60	M	Epoxide–CH_2–O–	2.65 3.15	GLYCIDYL ETHER	CDCl$_3$	139
3.60	SS	CH_3–O–C(=O)–	0.90 1.05 1.85	POLYMETHACRYLATE	CDCl$_3$	64
3.60	T	–O–CH_2–CH_2–O–C(=O)–	4.15	DIETHYLENE GLYCOL	CDCl$_3$	102
3.64	SS	–O–CH_2CH_2–O–	- - - -	POLYGLYCOL	CDCl$_3$	88
3.70	S	Cl–CH_2–CH–O–	- - - -	POLYEPOXIDE	CDCl$_3$	140
3.75	BM	CH_2–CH–O–	1.60 4.85	POLYVINYL FORMAL	ODCB	79
3.75	BM	CH_2–CH–O–	1.60 4.85	POLYVINYL FORMAL	CDCl$_3$	78
3.75	BM	CH–CH_2–O–C(=O)–	0.92 1.95	POLYMETHACRYLATE	CDCl$_3$	68
3.75	BB	$\emptyset$–CH_2–$\emptyset$	6.65	PHENOL–FORMALDEHYDE	C$_3$H$_6$O	151
3.88	S	$\emptyset$–CH_2–$\emptyset$	7.03	POLYISOCYANATE	CDCl$_3$	131
3.88	T	–S–CH_2–CH_2–O–	2.95	POLYSULFIDE	CDCl$_3$	97
3.90	BB	$\emptyset$–CH_2–$\emptyset$	7.2	DIISOCYANATE	CDCl$_3$	134
3.92	T	–S–CH_2–CH_2–O–	2.95 4.75	POLYSULFIDE	CDCl$_3$	97
3.95	BB	–O–$CH(CH_2)_2$	0.95 1.60 4.6	POLYVINYL BUTYRAL	CDCl$_3$	80
3.95	BB	CH_2–CH_2–N–C(=O)–	2.25 3.3	VINYL PYRROLIDONE	CDCl$_3$	69
3.95	BB	$\emptyset$–CH_2–$\emptyset$	7.21	DIISOCYANATE	TFA	132
3.96	T	R–CH_2–CH_2–O–C(=O)–	0.98 1.5 2.18	POLYACRYLATE	CCl$_4$	63
3.96	BM	R–CH_2–CH_2–O–C(=O)–	0.95 1.55 1.85	POLYMETHACRYLATE	CDCl$_3$	67
4.00	CM	CH–CH_2–O–	2.65 3.3	GLYCIDYL ETHER	CDCl$_3$	142
4.00	Q	CH_3–CH_2–O–C(=O)–	0.90 1.27 1.85	POLYMETHACRYLATE	CDCl$_3$	66
4.05	BB	$(CH_2)_2$–CH–OH	1.7	POLYVINYL ALCOHOL	D$_2$O	56
4.07	Q	CH_3–CH_2–O–C(=O)–	0.90 1.9	POLYACRYLATE	CDCl$_3$	62

CHEM SHIFT	BAND TYPE	STRUCTURE	OTHER BANDS	POLYMER TYPE	SOLVENT	PAGE
4.07	T	$-(CH_2)_3-CH_2-O-C(=O)-$	1.63	POLYESTER	$CDCl_3$	104
4.08	S	$R_3-C-CH_2-O-C(=O)-$	- - - -	POLYESTER	$CDCl_3$	185
4.08	T	$-CH_2-CH_2-O-C(=O)-$	1.25 1.84 4.95	POLYESTER	CCl_4	105
4.15	T	$-O-CH_2-CH_2-O-C(=O)-$	3.6	POLYESTER	$CDCl_3$	102
4.16	M	$-O-(CH_2)_3-CH_2-O-C(=O)-NH-$	7.05 7.27 7.67	URETHANE	CCl_4	136
4.28	SS	$-(O=)C-O-CH_2CH_2-O-C(=O)-$	- - - -	POLYESTER	$CDCl_3$	101
4.30	M	$-(CH_2)_3-CH_2-O-C(=O)-$	1.87	POLYESTER	TFA	133
4.45	BB	$-CH-CH_2-O-C(=O)-$	1.35 2.0	POLYESTER	TFA	109
4.48	BB	$-(CH_2)_2-CH-Cl$	2.14	VINYL CHLORIDE	ODCB	54
4.49	S	$-(O=)C-O-CH_2CH_2-O-C(=O)-$	- - - -	POLYESTER	TFA	132
4.60	BB	$CH_2-CH(-O-CH_2)_2$	0.95 1.6 3.95	VINYL BUTYRAL	$CDCl_3$	80
4.60	S	$\emptyset-CH_2-O-$	7.0	PHENOL-FORMALDEHYDE	$CDCl_3$	153
4.62	S	$R-O-CH_2-O-R$	1.73 2.78 3.52	POLYSULFIDE	$CDCl_3$	98
4.63	CM	$R-O-CH_2-O-R$	1.6 3.75	POLYVINYL FORMAL	$CDCl_3$	79
4.75	S	$R-O-CH_2-O-R$	2.95 3.91	POLYSULFIDE	$CDCl_3$	97
4.76	CM	$-CH-CH=CH_2$	1.25 1.78 5.05	POLYBUTADIENE	$CDCl_3$	37
4.85	CM	$R-O-CH_2-O-R$	1.60 3.75	POLYVINYL FORMAL	ODCB	78
4.88	SS	$CH_2-CH_2-O-C(=O)-\emptyset$	8.19	PET	TFA	108
4.88	BB	$(CH_2)_2-CH-O-C(=O)-$	1.80 2.02	POLYVINYL ACETATE	$CDCl_3$	61
4.95	M	$R_2-CH-O-$	1.25 1.84 4.08	POLYESTER	CCl_4	105
5.05	CM	$CH-CH-CH_2$	1.25 1.78 4.76	POLYBUTADIENE	$CDCl_3$	37
5.20	M	$-O-CH_2-O-$	5.4 5.65	POLYOXYMETHYLENE	TFA	87
5.25	BB	$-CH_2-CH-C$	1.75 2.12	POLYISOPRENE	$CDCl_3$	39
5.25	BB	$-CH_2-CH-C$	1.75 2.12	NATURAL RUBBER	ODCB	40

CHEM SHIFT	BAND TYPE	STRUCTURE	OTHER BANDS	POLYMER TYPE	SOLVENT	PAGE
5.40	SS	$-O-CH_2-O-$	5.2 5.65	POLYOXYMETHYLENE	TFA	87
5.40	T	$-CH_2-CH=CH-CH_2-$	2.09	POLYBUTADIENE	$CDCl_3$	35
5.65	SS	$-O-CH_2-O-$	5.2 5.4	POLYOXYMETHYLENE	TFA	87
6.35	BM	Pyridyl	1.82 6.79 7.12 7.69	POLYVINYL PYRIDINE	$CDCl_3$	70
6.50	S	Aryl	2.11	POLYPHENYLENE OXIDE	$CDCl_3$	95
6.60	BB	Phenyl	1.5 1.85 7.06	POLYSTYRENE	$CDCl_3$	42
6.60	BB	Phenyl	1.5 2.00 7.06	POLYSTYRENE	$CDCl_3$	43
6.65	BB	Aryl	3.75	PF RESIN	C_3H_6O	151
6.79	BM	Pyridyl	1.82 6.35 7.12 7.69	VINYL PYRIDINE	$CDCl_3$	70
6.90	S	$-(O=)C-CH=CH-C(=O)-$	- - - -	FUMARATE	$CDCl_3$	110
7.00	CM	Aryl	1.22	PF RESIN	$CDCl_3$	152
7.00	S	Aryl	1.60	BISPHENOL-A	$CDCl_3$	114
7.00	CM	Aryl	1.71 7.8	BISPHENOL-A	$CDCl_3$	96
7.00	CM	Aryl	0.70 1.31 1.69	OCTYLPHENOL	$CDCl_3$	154
7.03	M	Aryl	3.88	POLYISOCYANATE	$CDCl_3$	131
7.05	M	Aryl	4.16 7.27 7.67	TDI	CCl_4	136
7.06	BB	Phenyl	1.5 1.85 1.6	POLYSTYRENE	$CDCl_3$	42
7.06	BB	Phenyl	1.5 2.0 6.6	POLYSTYRENE	$CDCl_3$	43
7.10	S	Phenyl	1.6	STYRENE COPOLYMER	$CDCl_3$	45
7.10	BM	Pyridyl	1.82 6.35 6.79 7.69	VINYL PYRIDINE	$CDCl_3$	70
7.10	S	Aryl	1.65	BISPHENOL-A	$CDCl_3$	114
7.20	M	Aryl	3.9	MDI URETHANE	$CDCl_3$	134

CHEM SHIFT	BAND TYPE	STRUCTURE	OTHER BANDS	POLYMER TYPE	SOLVENT	PAGE
7.21	S	Aryl	4.49	MDI URETHANE	TFA	132
7.27	M	Aryl	4.16 7.05 7.67	TDI URETHANE	CCl_4	136
7.39	BB	Aryl	7.62	PHENYL SILICONE	$CDCl_3$	167
7.51	T	Aryl	8.22 8.68	ISOPHTHALATE	$CDCl_3$	110
7.62	BB	Aryl	7.39	PHENYL SILICONE	$CDCl_3$	167
7.67	M	Aryl	4.16 7.05 7.27	TDI URETHANE	CCl_4	136
7.69	BB	Pyridyl	1.82 6.35 6.79 7.12	VINYL PYRIDINE	$CDCl_3$	70
7.80	D	Aryl	1.71 7.0	POLYSULFONE	$CDCl_3$	96
8.06	SS	Aryl	- - - -	TEREPHTHALATE	$CDCl_3$	111
8.20	SS	Aryl	- - - -	TEREPHTHALATE	TFA	108
8.22	D	Aryl	7.51 8.68	ISOPHTHALATE	$CDCl_3$	110
8.30	BB	R—**NH**—C(=O)—R	1.5 1.85 2.7 3.5	NYLON 6, 6	TFA	119
8.68	S	Aryl	7.51 8.22	ISOPHTHALATE	$CDCl_3$	110

MANUFACTURERS INDEX

This index presents the trademark owners of the commercial names of the polymers and resins which have been cited as examples at the end of each chapter heading. The order in which the company names appear is based upon the order in which their trademarked products occur in the text.

This listing should not be considered a complete tabulation of major polymer and resin manufacturers. A listing such as that would be beyond the scope of this volume.

(1)	E. I. du Pont de Nemours & Co.
(2)	Dart Industries, Inc., Fiberfil Division
(3)	The Dow Chemical Co.
(4)	Phillips Petroleum Co.
(5)	Hercules, Inc.
(6)	Avisun Corporation
(7)	BASF Wyandotte Corp.
(8)	Enjay Chemical Co.
(9)	Imperial Chemical Industries Limited, Plastics Division
(10)	Uniroyal Chemical, Division of Uniroyal, Inc.
(11)	Ameripol, Inc.
(12)	Goodyear Tire & Rubber Co.
(13)	American Rubber and Chemical Co.
(14)	Firestone Synthetic Rubber & Latex Co.
(15)	Tex-Chem Co.
(16)	Koppers Co., Inc.
(17)	Monsanto Co.
(18)	Pennsylvania Industrial Chemical Corporation
(19)	United Carbon Co.
(20)	American Cyanamid Co., Industrial Chemicals & Plastics Division
(21)	Dow Badische Co.
(22)	BP Chemicals International Ltd.
(23)	B. F. Goodrich Chemical Co.
(24)	Farbwerke Hoechst AG

(25)	Scott Bader Co.
(26)	M & B Plastics Ltd.
(27)	May and Baker Ltd.
(28)	Wacker-Chemie GmbH
(29)	Badische Anilin- & Soda-Fabrik AG
(30)	W. R. Grace & Co., Polymers and Chemicals Division
(31)	Montacatini Edison S.p.A.
(32)	BASF Wyandotte Corp. (see also (7)).
(33)	Ashland Chemical Division, Ashland Oil Co.
(34)	Rohm & Haas Co.
(35)	Hardman and Holden Ltd.
(36)	GAF Corp.
(37)	Ionac Chemical Co.
(38)	Polymer Corp., Ltd. Canada
(39)	Shawnigan Resins Corp.
(40)	Celanese Plastics Co.
(41)	Union Carbide Corporation
(42)	Olin Chemicals
(43)	Farbenfabriken Bayer AG
(44)	Jefferson Chemical Co., Inc.
(45)	The Quaker Oats Co.
(46)	General Electric Co.
(47)	Westlake Plastics Co.
(48)	Diamond Shamrock Chemical Co.
(49)	Thiokol Chemical Corp.
(50)	Eastman Chemical Products, Inc.
(51)	Millhaven Fabrics Limited
(52)	Beaunit Fibers, Division of Beaunit Corp.
(53)	Mobay Chemical Co.
(54)	Allied Chemical Corp.
(55)	Gulf Oil Co.
(56)	Polypenco Ltd.
(57)	Aquitaine Chemicals, Inc.

(58)   Chemische Werke Huls AG

(59)   Emery Industries, Inc.

(60)   General Mills Chemicals, Inc.

(61)   Commercial Solvents Corp.

(62)   The Upjohn Co.

(63)   Tenneco Chemicals, Inc.

(64)   Thombert, Inc.

(65)   Stauffer Chemical Co.

(66)   ESB, Inc.

(67)   Ciba-Geigy Corp.

(68)   Celanese Resins, Division of Celanese Coatings Co.

(69)   Reichhold Chemicals, Inc.

(70)   Shell Chemical Co.

(71)   Hitachi Chemical Co.

(72)   Plastimer, S. A.

(73)   Kreidl, Rutter & Co.

(74)   Dynamit Nobel of America, Inc.

(75)   Adhesive Products Corp.

(76)   Emerson & Cuming, Inc.

(77)   Dow Corning Corp.

(78)   Allis Chalmers

(79)   UCB-Sidac

(80)   Bakelite Xylonite Ltd.

(81)   Cobon Plastics Corp.

(82)   British Celanese Ltd.

(83)   Neville Chemical Co.

(84)   FMC Corp.

(85)   Rexall Chemical Company

COMMERCIAL NAME INDEX

Name	Mfgr.	Page	Name	Mfgr.	Page
Acelon	(27)	170	Carbowax	(41)	86
Acetophane	(79)	170	Castethane	(62)	130
Acrilan	(17)	50	Celadex	(82)	170
Acrylite	(20)	50	Celcon	(40)	86
Adrub	(75)	160	Cellon	(74)	170
Alathon	(1)	24	Cellonex	(74)	170
Aldacol	(35)	51	Cipoviol	(26)	50
Alfane	(66)	138	Cisdene	(13)	25
Alloprene	(9)	51	Cis-4	(4)	25
Ameripol CB	(11)	25	CMC 120	(5)	18
Ameripol SN	(11)	25	Cobocell	(81)	170
Ameripol 4502	(11)	25	Coral	(14)	25
Ameripol 4503	(11)	25	Creslan	(1)	50
Aminolac	(72)	150	Cymel	(72)	150
Araldite	(67)	138			
Arnel	(40)	170	Dacron	(1)	100
Aroset	(33)	50	Dapon 35	(84)	18
Arylon	(10)	86	Daratak	(3)	50
			Darvic	(9)	50
Bakelite	(41)	18	Deligna	(73)	148
Baytown	(19)	25	Delrin	(1)	86
Beetle	(20)	150	Demilan	(73)	150
Bexoid	(80)	170	Densite	(63)	130
Breon	(22)	50	Deresit	(73)	148
Budene	(12)	25	Desmoflex	(43)	130
Budium	(1)	25	Desmophen	(43)	86
Butvar	(39)	51	Desurit	(73)	150
			Diakon	(9)	50
Cabulite	(27)	170	Diaron	(69)	150
Capran	(54)	117	Diene	(14)	25
Caprolan	(55)	117	Dion	(48)	86, 100
Carbamac	(61)	130	Duradene	(14)	25
Carbascar	(2)	100	Durethane	(43)	117

Name	Mfgr.	Page	Name	Mfgr.	Page
Paraplex	(34)	100	Styrofoam	(3)	25
Parlon	(5)	51	Styron	(3)	25
PEG 1000	(54)	18	Sylgard	(77)	160
Perspex	(9)	50	Synpol	(15)	25
Pevalon	(27)	50			
PE 204C	(85)	18	Terylene	(51)	100
Philprene	(4)	25	Tetronic	(32)	50
Piccolastic	(18)	25	Texicote	(25)	50
Pioloform B	(28)	51	Thanol F-3002	(44)	86
Pioloform F	(28)	51	Thermalux	(47)	86
Plexiglas	(34)	50	Thiokol	(49)	86
Plioflex	(12)	25	TPX	(9)	24
Pliolite latex VP-100	(12)	51	Tyrin	(3)	51
Pliovic	(12)	50			
Polectron	(36)	51	Uformite	(34)	150
Polidene	(25)	50	Unithane	(49)	130
Pollopas	(74)	150			
Poly G	(42)	86	Velon	(14)	50
Polymeg	(45)	86	Versalon	(60)	117
Polyox	(41)	86	Versamid	(60)	117
Polysar latex 781	(38)	51	Vestamid	(58)	117
Polyviol	(28)	50	Viclan	(9)	50
PVP K-30	(36)	18	Vinavil	(31)	50
			Vinnapas	(28)	50
Rilsan A	(57)	117	Viscasil	(46)	160
Rilsan B	(57)	117	Vistalon	(8)	24
Royalene	(10)	24	Vistanex	(8)	24
RTV	(46)	160	Vycron	(52)	100
Saran	(–)	50	Zefran	(21)	50
Silaneal	(77)	160	Zytel	(1)	117
Silaprene	(43)	160			
Silastic	(77)	160			
Silco-Flex	(78)	160			
Solprene	(4)	25			
SR-173	(46)	160			

ALPHABETICAL INDEX *

* The underlined numbers refer to the page on which a spectrogram of the polymer or resin is presented.